NO WISDOM WITHOUT FOLLY

THE EXTRAORDINARY LIFE OF FRANÇOIS ENGLERT, NOBEL LAUREATE

Chronicle in Ten Episodes, a Prologue and an Epilogue

NO WISDOM WITHOUT FOLLY

THE EXTRAORDINARY LIFE OF FRANÇOIS ENGLERT, NOBEL LAUREATE

Chronicle in Ten Episodes, a Prologue and an Epilogue

DANIELLE LOSMAN

Translated from French by Jane Kuntz

NEW JERSEY · LONDON · SINGAPORE · BEIJING · SHANGHAI · HONG KONG · TAIPEI · CHENNAI · TOKYO

Published by

World Scientific Publishing Co. Pte. Ltd.

5 Toh Tuck Link, Singapore 596224

USA office: 27 Warren Street, Suite 401-402, Hackensack, NJ 07601

UK office: 57 Shelton Street, Covent Garden, London WC2H 9HE

British Library Cataloguing-in-Publication Data
A catalogue record for this book is available from the British Library.

NO WISDOM WITHOUT FOLLY
The Extraordinary Life of François Englert, Nobel Laureate

ISBN 978-981-12-8324-6 (hardcover)
ISBN 978-981-12-8391-8 (paperback)
ISBN 978-981-12-8325-3 (ebook for institutions)
ISBN 978-981-12-8326-0 (ebook for individuals)

For any available supplementary material, please visit
https://www.worldscientific.com/worldscibooks/10.1142/13591#t=suppl

Typeset by Stallion Press
Email: enquiries@stallionpress.com

Printed in Singapore

For my daughter Judit and my grandson Sybren

François Englert undertook the labor of memory so essential to the realization of this text, an immense labor, both painful and salutary.

He dedicates this work to his parents, Szmul-Josek, called Joseph, and Baltche.

Their memory has forever been with him.

Foreword

The book you are holding is a rather special one.

Its title as expansive as an illumination, its table of contents a chest of drawers, its index foreshadowing a profusion of assorted characters. Paging through, you will notice inserts of various lengths and headings — intermezzo, "école buissonnière"[1]... You will happen upon a Feynman diagram, a poem, a Spinoza quotation, the story of a student prank, and a devastating passage about what it meant to be a "hidden child."

This is all about François Englert, whom you will discover as you read this book, and whom you cannot help but love.

And this has all been cleverly assembled by Danielle Losman, physicist and writer, who wished this biography to conserve the freshness and spontaneity of the countless interviews around which it was built. If readers leap from a gilded Stockholm palace to sleepless nights wracked by painful memories, if they move from a supremely lucid exposé on the "Brout–Englert–Higgs Mechanism" to discursive flights on determinism, if a moving reference to a deceased friend follows a hilarious account of a sputnik landing in Belgium, that's just "how life is."

This book is a Bruegel painting where unbridled characters overflowing with life dance to the music of a klezmer band.

---

[1] Literally, to skip school, to be truant. But it can also refer to the school of life, learning outside the official confines of an education system, off the beaten track.

It is an ode to the magical universe of physics, to intelligence and laughter, to friendship and resilience, to humanism.

Nathalie Deruelle
Emeritus director of Research at the CNRS

Notice

This narrative is an attempt at charting the itinerary of the physicist François Englert in both the world of science and in his day-to-day life, since his childhood until that day in 2013 when the Nobel Prize came to crown a lifetime's work. The genesis of this text involved numerous sessions during which François Englert sorted through his memories, recalling various episodes of his life. His professional work had its place, but we chose to treat that aspect with a "lighter" touch, so that the uninitiated reader might find enjoyment while discovering the exciting, wondrous world of scientific research.

As we move forward, we shall be defining, as simply as possible, the scientific notions whose intuitive understanding we feel readers will find useful as they follow the narrative, by deliberately leaving out any mathematical formulations.

A small chart listing notation and acronyms, a short bibliography, and a name index are provided as back matter.

Parenthetical indications are scattered throughout the text to refer to an episode and section where needed. (e.g., Section V.2).

The original texts of citations are provided with their references at the end of the volume. All the translations were done and/or checked by the author.

Contents

Notations

In the very few equations thought to be useful in this text:

- a point, or × (or nothing) represent a multiplication sign
- a forward slash is the division sign

Example:
$a.b = ab = a \times b = a$ multiplied by b
$a/b = a$ divided by b
$\approx$ the wavy equal sign means approximately equal to.

Exponents of 10

The exponent written after the symbol ^ represents the number of zeros after the 1.

A number whose exponent is negative is the inverse of the number whose exponent of the same value is positive.

Example:
$10^4 = 10{,}000 =$ ten thousand
$10^{(-4)} = 1/10{,}000 =$ one ten thousandth

Units

Energy: electron-volt (eV)

Frequency: Hertz (Hz)

Length: meter (m), centimeter (cm), light-year (AL)

Temperature: degree Celsius (°C), degree Kelvin (°K)

                 $0°K$ = absolute zero = $-273°C$

Time: second (s)

Unit exponents

cm^2 = square centimeter

cm^3 = cubic centimeter

Abbreviations and Symbols

c speed of light ≈ 3.10^10 cm/s

LY	= light-year
lambda λ	= wavelength
nu ν	= wave frequency
h	= Planck constant
p	= momentum

Acronyms

CERN — Conseil européen pour la Recherche nucléaire (European Organization for Nuclear Research)

CNRS — Centre national de la Recherche scientifique (France) (The French National Center for Scientific Research)

ENS — École normale supérieure (one of the highly competitive grandes écoles in France)

FNRS — Fonds national de la Recherche scientifique (Belgique) (The Belgian National Fund for Scientific Research)

INR — Institut national de Radiodiffusion 1930–1960 (forerunner of the RTBF)

KGB — Committee for State Security, USSR (1954–1991)

KUL/KU Leuven — Katholieke Universiteit Leuven

MIT — Massachusetts Institute of Technology

RTBF — Radio-Télévision-belge de la Communauté française (French-language radio/television company)

UCL — Université catholique de Louvain (Louvain-la-Neuve)

ULB — Université libre de Bruxelles

UMons — Université de Mons

Université NDP — Namur Université Notre-Dame de la Paix

VUB — Vrije Universiteit Brussel (sister university to the ULB)

Acknowledgments

The author wishes to thank friends who so generously read the manuscript and offered their advice and support: Nathalie Deruelle, Katelijne De Vuyst, Pierre Geron, Danielle Jacobs, Béatrice Lederer, Elizabeth Caplun, Jean Content, Albert Goldbeter, Nicky De Mayer, Marc Sapir, Guy Duplat, Mireille Soeur, Michel Cassé, and Marina Solvay.

François Englert gratefully acknowledges his friends Nathalie Deruelle and Philippe Spindel for the particularly enlightening discussions that nourished this chronicle.

The author also wishes to extend special thanks to François Englert for the many hours spent sharing his story, for his important scientific contribution, and his careful rereadings of the text, for his kindness and commitment throughout all the stages of this work over the long haul, and for familiarizing me, with both rigor and conviviality, with the subtleties of his scientific thought.

Prologue: A Physics Lesson

The Schrödinger Equation

October 1964, the Université libre de Bruxelles,[1] Lameere Auditorium, second-year Physics,[2] first lesson: Introduction to Quantum Mechanics.[3]

Lameere Auditorium is a lovely amphitheater, all in wood, a period style with art nouveau ironwork, large windows, a soft grey light with a few green accents, the reflection of trees out on Square Jean Servais.

In the middle of the green chalk board, the Schrödinger Equation, in all its glory.

Professor Englert paced the rostrum in a leisurely manner, after tossing his jacket, sweater, and tie onto the lectern. He didn't jump into explanations; he first wanted us to get a feel for what this basic equation meant,

---

[1] L'Université libre de Bruxelles, or The Free University of Brussels (ULB), founded in 1834 on the principle of free inquiry, "for the purpose of combating intolerance and prejudice, and to spread the philosophy of the Enlightenment."

[2] The Belgian university curriculum generally includes two years of *candidatures*, roughly equivalent to an associate degree, then two or three years of *licences*, a full graduate degree or master's.

[3] Quantum mechanics, developed during the first two decades of the 20th century, analyzes systems (such as atoms and their constituent parts) whose dimensions are smaller than one one-hundred-millionth of a centimeter and which do not obey the so-called classical laws of physics. Today, we prefer the more general term of "quantum physics." We will present the founding concepts of quantum physics in Section III.2. The Schrödinger Equation describes in quantum physics the evolution of a physical system.

appealing to our intuition and imagination. His eyes sparkled, his fine hands caressed the words that fluttered about in his cigarette smoke. He extrapolated elatedly.

He didn't seem so different from the man in the sunny yellow scarf that we sometimes ran into at Beppino's, a tiny, dimly lit eatery frequented by young ULB profs and assistants — known for its caged canary, its imperious owner, Madame, and the best coffee in town. He would show up, crackling with mischievous fun, a man who knew by heart the whole Brassens songbook and could sing Brecht in the bleakest tones:

Erst Kommt das Fressen
Dann kommt die Moral.[4]

But on that day, before this chalkboard covered in mysterious signs, his voice delivered a world of beauty and harmony. The elegance of mathematics with, as a bonus, the thrill of knowing that the secrets of nature were soon to be revealed.

After class, Professor Englert returned to his office in the massive red-brick Faculty of Sciences building, lost in thought.

He took the underground passageway — those long subterranean corridors leading to the science building.

Anyone who made regular use of these hallways knew the grain of every wall, the peculiar smell of each segment, the sound of footsteps on the tile; the low ceilings crisscrossed by large-gauge electrical conduits that disappeared into the darkness, nooks and crannies full of surprises: an American researcher tying his sneakers, a malingering facilities worker, a flunked student contemplating some revenge... Art Nouveau spiral staircases led to small laboratories that were wedged in at some point when there wasn't enough room on the more "noble" upper floors.

People tended to rush through these hallways, between classes.

But Professor Englert never broke his stroll.

He went up to his office: bright-colored curtains, an orange sofa. His chalkboard was covered in even stranger signs. He lit a cigarette that he smoked from a cigarette holder. A way to stop smoking, since cigarette

---

[4] "First eat, then moralize."

holders make tobacco taste awful, but "as I'm a masochist like everyone else, I smoke even more!"

A relaxing moment. He was smiling. Perspiration made his dark hair curl.

Who is this man who exudes such intelligence and joie de vivre? Some claim we can't be both intelligent and happy. So, does this joy conceal a dark soul, a vulnerability, and is his good humor, as Chris Marker said (and not Boris Vian), the politeness of despair?

Who is François Englert?

Among my memories as a former student, images jostle: there was the young professor whose contagious enthusiasm initiated students to the charms of quantum mechanics, the one we saw wandering around in winter, bundled up in a dark pea coat and long yellow knit scarf, his head buried in a sumptuous fur hat, and in the summer, wearing a pullover with holes at the elbows over corduroy trousers. There was the laidback science nerd with the caustic sense of humor whom we heard in constant conversation with his philosopher friend Georges Miedzianagora, before going off to "serve up some nonsense," by which he meant, go teach a class. The lively guest at a little dinner party, but also the troubled witness to a troubling world, who, despite his deep skepticism, did not hesitate to move beyond the passive role of bystander when he sensed the need.

As I got to know him better, from my student days and throughout all the subsequent years, for we have maintained a loose friendship, revived whenever we chance to meet, the predominant image was that of a rebel, a man with a deeply rooted urge to challenge, one who refused to fit into any mold imposed by the customs and traditions of any milieu, who asserted a total freedom to act according to his convictions or to a moment's inspiration, and who made it known bluntly, resorting to provocation when the situation required it. A man for whom independence was a person's most priceless asset. I would understand later that this essential feature of his character appeared very early: from the schoolboy who signed his own report cards, preferring that his parents not meddle in his school business, to the adolescent who, finding the world of adults disconcerting, even hostile, swore one day that he would never turn out like them.

And throughout his life brimming with friendships, passions, challenges, anything but "a bed of roses," he would stay the course, always the sole master and commander.

Is this spirit of contestation integral to his great intelligence, his unquenchable thirst to decipher the universe, or is it a consequence of all that, perhaps a corollary?

And in this equation, what role did family legacy play, how much did his childhood figure? What seedbed allowed this rebellious life to germinate?

This book is an attempt at answering these questions.

Episode I

A Child in Wartime

My writing has sprung from the seedbed made up of everything that happened to me in my childhood and early adolescence, and if I were to include only later experiences, without the primal home — from the foundations to the rooftop — I would be drowning in a sea of contradictory reflections. [...] Creation is always linked to the mysterious gaze of the inner child that leaves an imprint that cannot be transformed by any literary ruse.

Aaron Appelfeld, Mon père et ma mère. De l'Olivier 2020
(Translation from the French)

I.1. The Birth of a Project

It was after the awarding of the Nobel Prize in Physics in 2013, and quite by chance, that the project of recounting a few key moments of François Englert's life came to be: a suggestion by some friends of François's who proposed it during a conversation. The idea immediately appealed to me, which is how I found myself one day at his home, notebook in hand, settled into a comfy, beige leather armchair.

He set out to tell me the story of his life, starting naturally with his childhood. A story much restrained in the telling: facts and anecdotes, yes, but of his private experience, he said practically nothing. On rare occasions, a darkening of the eyes, a barely perceptible crack in his voice, betrayed the emotion. One intuits how much is yet unspoken in the recollection of this Jewish childhood when, at age seven and a half, he experienced the outbreak of World War II.

I had to come up with the right questions to break through the wall of silence that continued to obliterate the memory of so many survivors of the Shoah.[1]

He talked and I jotted down notes. Then, based on these notes, I wrote. After childhood came adolescence, school, and friends — a tangle of memories and anecdotes. But still this persistent restraint. François refrained from showing any emotion. However, when we moved on to the subject of his growing interest in mathematics first and in the sciences later on, his enthusiasm was unleashed. He took obvious pleasure in detailing his early avidness, in sharing his young researcher's excitement. The university setting that he discovered was described lucidly, no holds barred. But despite his good-humored skewering of the way the institution was run, François retained his soft spot for humankind.

As we pursued our conversations, François, who had never before considered telling his own story, was confronted with his life, with his past. He was compelled to piece together episodes, explain choices, and recall his motivations. Buried memories resurfaced. The contrast between the somewhat chaotic pattern of his affective life, on the one hand, and the luminous path of his life as a physicist, on the other, gave him pause. He was starting to get excited about this reconstruction puzzle.

But all of a sudden, something was wrong. He tells me over the phone that there is a problem, and we decide to get together as soon as possible.

It is his wife Mira who greets me. She seems very emotional.

"The work that you are doing has got him thinking quite a bit, about his childhood especially. He almost never talks about it, but last night, he opened up. You know that François was a 'hidden child.'[2] His childhood,

---

[1] The Shoah: A Hebrew word that means "catastrophe," referring to the systematic extermination of Jews carried out by Nazi Germany during the Second World War.

[2] Term commonly used to designate Jewish children who were hidden during the Second World War by people in occupied countries willing to take them in at risk to their own lives.

he has kept it to himself for the longest time, but it's always with him, always present."

François comes to join us.

He explains:

"If you manage to write down the memories I recount, fine, but if that's all there is to it, what's the point?"

For him, this book has to try to answer the questions he had been asking himself about life, the ones we all ask eventually: Why have I lived as I have? Why did I become what I am? And as is the case for most, many answers are to be found in childhood. Son of immigrants, hidden child, François was a child who survived. To understand his life meant to first attempt to understand how survivors managed to carry on living.

Intermezzo 1: Children of the Shoah

One million five hundred thousand Jewish children were murdered in the death camps.[3]

A child had no chance of surviving in the camps. Apart from some extremely rare exceptions, they were put to death upon arrival, and along with them, the women and men considered too weak to work.[4]

---

[3]The camps: from the earliest days of the war, the Nazis organized concentration camps in Poland to group the Jews together. Among them, six would become extermination camps, intended exclusively for the mass execution of human beings, an overwhelming majority of whom were Jews: Auschwitz-Birkenau, Bełżec, Treblinka, Madjanek, Sobibór, and Chełmno. Dozens of concentration camps (for internment, transit, and aggregation) were also created in Germany, as early as 1933, and in the occupied zone.

[4]"As soon as they arrived at the Auschwitz station, children under the age of fourteen were grouped together with the elderly and most of the women, and were taken to Birkenau, to the gas chamber. No sorting took place. This is what happened with the children's convoys coming from Drancy [outside Paris]. These convoys consisted of very small children, accompanied by a few nurses or social workers who were already working in the children's homes where they were at the time of their arrest." (Testimony of a former SS officer, CAIRN.Info, History of the Shoah).

The children who survived Nazi barbarity were those who managed to avoid the camps altogether. Some ten thousand children, nationals under the German Reich, were taken into care, starting in 1938 by relief organizations supported by the British government. Separated from their families, they were sent by convoy — the *Kindertransport*[5] — to Great Britain and to neighboring countries (which no one imagined would soon be invaded). Meanwhile, between 1935 and 1945, some one thousand children were sent over to the United States thanks to individual initiatives and a few private organizations. There were also those who, whether emigrants or native born, were hidden within the occupied countries themselves. These children were saved thanks to highly risky measures taken by their parents driven by a heroic selflessness and the courageous devotion of ordinary citizens. Still others, far fewer in this case, managed on their own — some even succeeded in joining one or another family member who had immigrated to the United States.

All these children experienced the crushing misfortune of being all at once torn from their families without really understanding why. They saw their world collapse from one day to the next, found themselves suddenly looked upon as pariahs, as hunted quarry, having to suppress their own names, to lie, wondering anxiously what had become of their family and relatives. Many landed among strangers unable to speak the local language, while others were subjected to humiliation and abuse.

All underwent major trauma.

All are "children of the Shoah."

For a long time, the adult survivors of this tragedy kept silent, and even those close to them were reluctant to ask questions. A law of

---

[5]The *Kindertransport*, (transport of children) is the name of a series of rescue operations between 1938 and 1940 that allowed for the transfer of thousands of child refugees from Nazi Germany to Great Britain. After the anti-Semitic violence organized by the Nazi authorities during *Kristallnacht* in November 1938, the British government relaxed their immigration restrictions for certain categories of Jews. Under pressure by British public opinion and refugee relief committees, particularly the British Committee for the Jews of Germany and the Movement for the Care of Children from Germany, the authorities consented to accepting an indeterminate number of children under the age of 17 coming from Germany and territories occupied by Germany (Austria and Czech territories) into Great Britain.

silence seemed to reign over the survivor experience, as if the Shoah had clouded the minds of a whole generation. However, sociologists, historians, and psychologists of all persuasions finally looked into the issue of adult survivors of the Shoah, and men and women finally emerged from their silence. We began to find out about what they went through and the strategies they adopted, consciously or not, to continue living.

Paradoxically, child survivors received far less attention, on the assumption that children adapt quickly to new situations, whatever they might be, and possess a vitality that impels them toward the future and allows them to forget the past.

One might well believe this, since the outcomes of these children, upon examination, demonstrated that most of them successfully reintegrated into society; indeed, a remarkable percentage of these men and women went on to achieve excellence, a far greater share than that of a comparable sample of the general population, and particularly in the sciences. Several Nobel Prizes have been awarded to former "children who survived." And this was but the tip of the iceberg. Many survivors contributed impressively[6] to the scientific and cultural progress of our civilization. Why and how? What was these children's itinerary? How did they live? What price did they pay?

For, although children are perhaps more resilient than adults when confronting major traumas, and many of these Jewish children did prove to be extraordinarily resilient, the price they paid needs to be taken into account. After decades of silence, these "children of the Shoah" started speaking out about the heavy psychological toll they suffered with each professional success, each victory over adversity. Even when these children "succeeded" in their adult life, a darker side always remained. There were no exceptions, there was no getting over the Shoah. But there are ways of grappling with it. Could professional excellence and success be a kind of revenge?

And yet, for many of them, success did not seem tinged with revenge. While it was important to face the enemy and declare, *You didn't get me, I'm still alive,* a motivation of a different order may well

---

[6] See Bibliography.

explain this drive to achieve: we know how much the Jewish communities of central Europe fostered respect for knowledge, love of culture, and how they wished to contribute to human progress. These are values handed down over generations that no catastrophe has yet succeeded to extinguish.

In that respect, François's itinerary is exemplary.

For him, this book, first and foremost, must be a hymn to the resilience of Jewish children.

They wanted to bury us, but they did not know that we were seeds.

François finds this Mexican proverb magnificent.

Yes, that's exactly it!

"The trauma they experienced became a vector of strength, of combativeness?"

No, not really…

"It's more complicated than that, I guess?"

No, it's not more complicated, but more subtle. You see, to understand something that's complicated, you need knowledge, but to understand what is subtle, you need to think.

This answer plunges me into the depths of thought.

Yes, each resilience was unique. For François, his frenzied need for independence, his visceral non-conformity, his perpetual questioning, were these all signs of his own resilience?

Will the story of his childhood provide us with the common thread of his life path, allowing us to go forth with him through the maze of his life, to understand his choices?

Perhaps. But despite all the good will François shows in giving his account, we will see that his childhood remains for him a land where he is still fumbling in the dark. The child he once was seems indeed remote, almost a stranger, and so different from the carefree children he interacted with since.

And then, as he himself says, it's much more subtle than that.

François, subtlety, Einstein, and God …

I notice that François fancies this word "subtle," often found in the writings of mathematicians and physicists.

He explains to me that the adage "Everything is always simpler than you think" is misleading, for in physics, everything is always subtler than you think.

Like an echo of that famous line introduced by Einstein in 1921, *Subtle is the Lord God, but malicious He is not.*[7]

This declaration appears doubly subtle when we know what Einstein wrote in January 1954 to his friend Erik Gutkind: *The word God is for me nothing more than the expression and product of human weakness, the Bible a collection of honorable, but still purely primitive, legends. No interpretation, no matter how subtle, could make me change my mind.*

And on 24 March of the same year, he replied to Joseph Dispentiere, an Italian émigré to New Jersey, a professed atheist who was concerned after reading in the news that Einstein was a believer:

It was, of course, a lie what you read about my religious convictions, a lie which is being systematically repeated. I do not believe in a personal God and I have never denied this but have expressed it clearly. If something is in me which can be called religious then it is the unbounded admiration for the structure of the world so far as our science can reveal it.[8]

Herr Gott, the Lord God — a nod to Nature, perhaps to the God of Spinoza,[9] that philosopher whom Einstein so admired.

I.2. Emigrants

In 1924, fleeing poverty and the virulent anti-Semitism rampant in Poland,[10] Szmul-Josek, called Joseph, and Bajla Englert left the small

---

[7] It is Oswald Veblen, a professor at Princeton, who heard Einstein make this statement during a discussion about an experiment that seemed to invalidate the theory of relativity. In 1930, Veblen wrote to Einstein to ask his permission to etch this declaration onto the lintel over the fireplace in the Math department, and Einstein gave his consent!

[8] https://www.library.ucsb.edu/research/db/1213.

[9] Baruch Spinoza (1632–1677), Dutch philosopher of Portuguese Jewish origin, one of the principal representatives of rationalism.

[10] Between 1921 and 1925, 400,000 Polish Jews emigrated to the United States, Palestine, Argentina, and Western Europe.

town of Radomsko in Poland with their son Marc, eight months old at the time.

They left behind their shtetl,[11] a world where life was hard, but it was their world, their language, their traditions, their friends, their memories. They did not suspect that this world was about to vanish. That a fanatical, bloody dictatorship would attempt to erase them from the face of the earth, and would nearly succeed. Three million Polish Jews were murdered by the Nazis. Out of the some 7,800 Jews living in Radomsko before the war, only 150 survived.

Joseph and Bajla Englert chose Belgium as their destination because two of Joseph's brothers, Uncles Ignace and Leon (Leibele for those close), lived there already.

Uncles Ignace and Leon ran a fabric store in Brussels. They got along well, lived together in Woluwé-St-Lambert, Avenue de Broqueville. They owned an automobile that only Leon ever learned how to drive. He was also the only brother who had served in the Polish military, which endowed him with a reputation as a fearless athlete above reproach. They were big personalities, extroverts, steeped in Yiddish humor. They were not observant, but were fiercely nostalgic for Yiddishland, while equally imbued with the most up-to-date political ideas, for they identified as Marxists. At any rate, they loved to taunt the bourgeoisie. When Uncle Ignace would dine out at the Chalet de la Forêt, a stylish restaurant located on the Drève de Lorraine in Uccle,[12] he took great delight in displaying his copy of Le Drapeau Rouge.[13] At the store, located on Sainctelette Square, where they would go every day on foot — quite a hike from Woluwé — he would charm the clientele. When a customer would ask, "Are you one of the Englert brothers?" he would come back with "Yes, but I'm the good-looking one."

---

[11] Yiddish word meaning little town, village, or neighborhood where the Jewish communities of Eastern Europe once lived, in a vast region dubbed Yiddishland.

[12] Uccle is one of the 19 districts, or *communes*, of Brussels.

[13] Daily newspaper of the Belgian Communist Party (1921–1991). Publication resumed in 2004 in the form of a bimonthly.

He would marry Fanny, a young well-educated woman, one of the sisters of Charles Krivine, a major fabric wholesaler of Russian origin who lived in Brussels since 1928.

Charles Krivine would play a certain role during the war with the relief committees for the families of Jews mandated to perform "forced labor abroad." When it became clear that "forced labor abroad" really meant deportation to concentration camps, Krivine approached the Belgian authorities, even King Leopold III himself, in an attempt to save members of his family, with little success. In fact, he and his wife only managed to survive by going underground.

The two uncles were also regulars at the Grand Veneur, the famous Keerbergen hotel-restaurant, where Leon would enjoy fascinating the proper young Flemish girls with his card tricks (and later, he would occasionally call upon François as his sidekick).

Szmul-Josek, called Joseph, was quite unlike his brothers. He was a discreet man, thoughtful, and somewhat introverted. Just as unobservant as his brothers, he was nevertheless deeply familiar with the sacred texts. Community members would sometimes even consult him on philosophical matters, or simply to discuss personal problems. Had he perhaps attended a yeshiva[14] back in Poland?

But Joseph also had quite the sense of humor, as we shall see.

Bajla, who was called Baltche, was proud, though unostentatiously so, of her husband; she was smart, lively, hardworking, and utterly devoted to her family. And dauntless too, for it was often thanks to Baltche's resourcefulness that the family managed to escape the worst during the war. Like her husband and brothers-in-law, she also sorely missed the world they had left behind. However determined they might have been to adapt to their new circumstances, their home life, like that of most immigrants, was full of little habits and features of their native shtetl existence.

Upon arriving in Belgium, the family lived in modest lodgings in Anderlecht. They then moved into the building where the fabric store

---

[14] Educational institution for the intensive study of sacred Jewish texts (the Talmud, the Torah, the Halakha).

was located, on Rue Jules Van Praet, near La Bourse. Baltche worked at the store, took care of the household, and kept things organized. She wasn't shy about sharing her opinions, all in a delightfully redolent language; she loved to nickname people, and everyone who worked at the store would soon earn a sobriquet that harkened back to Yiddishland. Blinde Beer, Djabbele, Pisherou, Shmoy Boy, and Imbeciel — for an especially ungifted salesman.[15]

Not unlike Joseph and his brothers, she loved to playfully tease. At François's second wedding, where the in-laws, staunchly bourgeois Catholics, were holding the reception at the grand Moulin de Lindekemale restaurant, Baltche declared, after tasting the soup, "Ah, now these are really some fine cans!" And Joseph piled on with, "You can tell it's a chic restaurant, there's almost nothing on the plate!"

In the thirties, the family moved once again, into a building located at the corner of Boulevard d'Ypres and de Dixmude: a small apartment, living and dining rooms, kitchen and bath, two railroad-flat bedrooms (the children had to go through the parents' room to get to theirs), an upper room for the Polish housemaid, and even a buzzer to call her from the dining room. This particular move signaled a certain upward social mobility — "we don't live above the store anymore." In fact, the store had also moved to Rue Léon Lepage, in the Dansaert neighborhood, near a now-defunct café, Le Vieux Rempart.

The family moved in on a Sunday, not realizing that on every other day of the week, starting at 4 in the morning, the pandemonium of a vegetable market would break out right beneath their windows. No, they hadn't achieved "the good life" just yet!

François was born on 6 November 1932.

Baltche was quick to nickname him Fajwel, Fajwel Drekele, or "Pavel, my little poo," but his birth certificate declared him to be François, a less exotic name, apparently inscribed there by some mysterious lady (a local administrator, perhaps). At any rate, François would become the customary name within the family.

Baltche was a devoted wife and mother, but as always with the Englerts, unostentatious. She loved her children, but was not demonstrative about it.

---

[15] Blinde Beer = blind bear; Djabbele = little frog; Pisherou = little pisser, etc.

François has no recollection of her coming to tuck him into bed. Joseph, on the other hand, would sometimes kiss his children goodnight. A tender father, a pragmatic mother? Perhaps.

Whatever the case, raised by a strong-willed mother and a somewhat withdrawn father, François grew convinced that it was women who were running the world, and that men, on the whole, played a secondary role. For a long time, he would manifest a certain shyness toward women.

But what was day-to-day life like in the Englert household?

The parents worked a great deal, spending most of their time at the store. François and Marc sometimes felt neglected, not understanding that all the parents wanted for them was a better life, one where they would be well integrated into Belgian life, where they might get educated and thrive. Even so, François felt very alone. Marc, eight years his senior, was not a real playmate. Of course, they would occasionally talk about this and that, communication between brothers being in any case easier than with parents, who did seem to be living their "separate life." François and Marc would later regret not having had more understanding of their parents' attitude. As François later put it, a half-century on:

> *As I grow old, I have better realized everything my parents did for me, during the war of course, but also in daily life. I've understood how much they sacrificed, how much love and trust it took to ensure that their children got well-integrated into the culture of a world that was not theirs, a world whose subtlety and humanity I came to appreciate.*[16]

Joseph and Baltche spoke to each other in Yiddish and in Polish to the maid. The father learned French and was soon fluent; he loved reading, and soon newspapers and books in French were part of the household. Baltche got by well enough. The children understood Yiddish but didn't

---

[16] Speech delivered to the Belgian House of Representatives on 23 January 2019, (Ceremony commemorating the "Journée internationale en mémoire des victimes de la Shoah").

speak it. The "mamaloshen," Mama's language, was not François's mother tongue; with maman, they spoke in French, somehow or other. But exiled as she was, Baltche must have sometimes felt helpless when it came to comforting her son when he needed to confide in his mother, as sons often do. And did François ever feel equally helpless? Might this somehow explain his characteristic reluctance to open up, and the resulting sense of isolation?

Family life was organized around work and school more than by the ritual calendar. The parents rarely attended synagogue, with the possible exception of Yom Kippur.[17] As for François, he never set foot there. The father tried his best to maintain certain traditions, especially Seder on the eve of Pesach,[18] though both boys took pleasure in undermining his efforts. (As we will see, François took this refusal of family tradition to an extreme.)

As was the case with many Jewish families in Poland, birthdays were not celebrated at the Englerts', nor were there any other celebrations marked by their family. We should remember that children were not showered with toys back then in quite the way they are today. A large top and, later on, an erector set, are the only toys François recalls ever receiving.

Nor did the family left behind in Poland ever come up in conversation. A paternal grandfather did come to Brussels for a short visit, but François has no memory of it.

Despite the relative comfort they managed to attain, Joseph and Baltche remained émigrés, exiles torn between the world they lost and their new life in a foreign world, strangers in a strange land. Nostalgia for the lost country lingered; they recreated the atmosphere around a table with friends, all Jews like them, where they would speak in Yiddish, talk about the past, play cards, where *vitsen*[19] would rekindle the flame of Jewish humor, their most treasured asset. The two brothers did not take part in these conversations, but its music remained in their ears, and an emotional François would later recall, *We weren't religious at home,*

---

[17] Yom Kippur is known as the Day of Atonement.

[18] Seder is the ritual meal on the eve of Passover.

[19] Jewish jokes.

but there was something... very Jewish. The coarse melody of Yiddish, the laughter, the laments, *oy vey iz mir.*[20] This "little tune" of nostalgia was to be the only family narrative François would carry with him throughout life.

In the end, I don't know much about my family, I don't know where I come from. People like to talk about their roots; me, I have the curious impression that I'm floating, he often told me. And each time he did, I felt we were touching on something both essential and imperceptible.

Though François says that he has few recollections from the pre-war period, one very precise memory remains: the loneliness that dominated the landscape of his childhood. But he worked hard to keep it at bay: in the bedroom he shared with Marc, he set up a little desk with everything he needed: a blackboard and chalk to invent and draw histories, stories of battles involving the heroes he most identified with, a whole universe.

When the time came for him to start kindergarten, Rue Locquenghien, he welcomed the new distraction, with all the activities that interested him. He had a thirst for learning, and was immediately fascinated by numbers and calculations. One morning, he and his mother arrived to find the door closed — no kindergarten on that day. François burst into tears. But this love for the school system would not last. His school years would prove exasperating, for he knew how to read even before entering his first year of primary school, having learned to sound out the advertisements on the tramway; he would end up learning them by heart. He even learned the rudiments of stenography from his brother Marc!

Reading became his pastime of choice, and would remain one of his favorite hobbies throughout the rest of his life.

A child his own age occasionally came over to the house. A boy named Hansi Rosenblatt, a young Austrian Jew, came to Belgium in 1939 thanks to the *Kindertransport*, the rescue operation I mentioned earlier, which, between 1938 and 1940, managed to get thousands of Jewish children out of Nazi Germany and Austria. Hansi was taken in

---

[20] Oh, woe is me!

by a Jewish family who were friends of the Englerts. The two boys would lose touch during the war, but met up again years later at some mutual friend's house. Little Hansi, formerly a "hidden child" like François, had grown to become Henri Roanne-Rosenblatt, renowned journalist, novelist, and filmmaker. Henri and François resumed the course of their companionship interrupted by the war, and a deep friendship would bind Henri and his wife Gladys with François and Mira. In 2019, they participated together in a commemoration of the *Kindertransport* held in Cologne.

For François, the time came to start elementary school, first grade, at Rue du Canal. Everything moved far too slowly for his liking. One day, the teacher gave the pupils a problem to solve at home, a math problem in a rectangle. François was quick to find the solution, all by himself, following his intuition, and he was the only pupil to have the answer. Later, he would recall this incident as the first time he experienced the thrill of resolving a scientific enigma!

Because classwork bored him, he was moved up one year, despite his mother's fears, and in 1939–1940, little François found himself in third grade.

He was a solitary pupil, having little to do with his fellow classmates, who mostly ignored him in turn. At home, he kept to himself, saying little — his school adventures and misadventures were nobody's business but his. Between his very busy parents and a rather dreary school life, his pre-war childhood unfolded without major incident, without sadness either, though it undoubtedly lacked the usual carefree effusiveness and the chatter. But despite it all, the family apartment was the place where François felt most at home, sheltered from the outside world, the place where he could let his imagination wander, lulled by the little music of conversation around the table, blending Yiddish and a hybrid of French.

Papa, Mama, brother Marc, and little François.

I.3. The War

On 10 May 1940, the whole family woke to the noise of the DCA.[21]

Around 6 am, Joseph burst into the room shared by François and Marc: "It's war! The Germans have attacked Belgium!"

After a moment of disbelief, the children understood and fear took root. François promptly threw up. It was at this precise moment, François tells me, that he began collecting memories.

There was panic in the streets. Alarming news was circulating. The German army was indeed invading Belgium. The thunder of cannons could be heard coming from Louvain!

---

[21] Anti-aircraft defense.

Three days later, the Englert family was on the move. There was a major exodus, with more than one million Belgians fleeing the country to seek refuge in France. For the Englerts, the chaotic trek would take several weeks; parents and children made up a small convoy that included the Englert and Krivine families. Those in Uncle Ignace's car driven by Uncle Leon headed to Lille to stay with paternal Uncle Sroultche, after which they pressed on to Tours, where Joseph's sister Esther lived — she was married to a Frenchman — and then on to Arcachon, before finally arriving in Toulouse, in the unoccupied zone where thousands of Belgian, Dutch, Luxembourger, and French refugees jumbled together.

Word was spreading among the refugees that the situation in Belgium was getting back to normal, so the family decided to return to Brussels. Baltche, listening only to her bravery, set out as scout, an agonizing separation, for there was no communication between the free zone and occupied Belgium. A few weeks later, thanks to a mail "detour" via Switzerland, Baltche managed to get a message through, that "all is quiet in Brussels," and that the family could go home.

A huge relief for all.

François went back to school. During the two first years of the war, daily life at the Englerts' returned to its normal rhythms, more or less, despite the discriminatory measures imposed by the occupiers.

On December 1st 1941, a German ordinance banned Jewish children from attending non-Jewish schools. Pupils for whom schooling was not compulsory were to leave their schools by 31 December.

With regard to Jewish pupils for whom school was compulsory, it was up to the Ministry of Public Education to determine the cutoff date, after which they would have to attend a school for Jewish children, which had not yet been created, for at that time, there were only two Jewish schools in Belgium, both in Antwerpen.

Immediate enforcement of the ordinance was therefore impossible. Besides, in Brussels, as in many other Belgian communes, the authorities did not hide the fact that they were only halfheartedly complying with these segregation measures. It turned out that during the 1941–1942 school year, most Jewish children in primary school in Brussels were able to attend the same institution they always attended.

By May 1942, it was compulsory for Jews to wear the yellow star.

But at the school on Rue du Canal, anti-Semitism did not pervade. On the contrary, certain teachers showed great kindness toward Jewish pupils, and sought to provide them with moral support. As for François, he was involved in a couple of little incidents.

One day, on his way home from school with a Flemish classmate who lived on the same street, and whose parents, it was rumored, were members of the VNV,[22] François was gripped by a sudden impulse and called him a collaborator. Upon this, the little Flemish boy socked him in the jaw. It hurt, but the next day, François had the satisfaction of seeing the boy arrive at school with his hand wrapped in a bandage!

At the Place Sainctelette Fair, at the bumper-car ride, François noticed that certain kids were aiming their cars at him in particular. Despite these episodes, he was not overwhelmed by any sense of fear, but he was henceforth on his guard. He was starting to realize that being Jewish was a rather special state, one that entailed risks.

The 1941–1942 school year drew to a close.

François recalls it with quite a bit of pride: on awards day, many yellow stars figured among the prize recipients! And something occurred to him that day: a certain awareness of his Jewishness. Never before had he cared about his difference, about his status as a Jewish child of parents who had come from afar, fleeing harassment, oppression, and hatred. He realized he would have to come to terms with certain things, still a vague sensation, but one he was ready to face.

At home, no one spoke about the "situation." The parents provided no explanation, and everyone acted as if nothing was the matter, but the tension was obvious. Of the very bad news that the parents must have been receiving from Poland, nothing filtered down to François, or at least, he has no memory of it.

---

[22]Vlaams Nationaal Verbond, the "National Flemish League," the Flemish Nazi party founded in 1933.

The store, like all "Jewish assets" in the conquered country, was put into custodianship, depriving the parents of any professional activity.

The noose was tightening.

Marc, who was attending secondary school at Rue des Riches–Claires, was expelled in January 1942. Jewish students over 16 years of age had to leave their schools to be grouped together in a school in Liège whose purpose was to train teachers for "future Jewish schools" that, of course, would never come into being.

In June 1942, Marc was summoned, as a Jew, to perform forced labor in Germany. What they did not know at the time was that this actually meant deportation to concentration camps in Poland.[23] But how could they not be suspicious? Joseph was devastated, Baltche very reluctant. To disobey would involve huge risks, while separation was even riskier. Marc's bag was packed, but at the last minute, Baltche refused to let him go.

The situation was clearly becoming very dangerous for all Jews.

I.4. The Hidden Child

Joseph and Baltche had contacts within the Resistance, through a nephew of Aunt Fanny, Uncle Ignace's wife. They forged some fake IDs for them, which Joseph had to go collect at the café "Le Vieux Rampart" on Rue Dansaert. But right before he entered, he noticed a Gestapo vehicle and beat a hasty but discreet retreat.

The family then made an eleventh-hour exit, in July 1942, without the forged ID papers, and just before the first round-ups.[24]

They left for Lustin, in the province of Namur, on the advice of Aunt Fanny's nephew, who knew Lustin well for having vacationed there.

They were housed at the Café-Restaurant de la Gare by the family of the restaurant owners, the Jourdans, who welcomed them with great

---

[23] See I, Intermezzo 1.

[24] The round-ups during the occupation consisted of mass arrests of Jews by the SS, often in collaboration with local police forces. To ensure the most successful outcome, the organizers of these dragnet operations relied on the element of surprise (sudden cordoning of streets, striking at dawn, bursting suddenly into homes without warning).

generosity and immense courage, since hiding Jews had become extremely dangerous.

The parents understood that they could not remain at the Café-Restaurant. They decided that it would be better if they separated, and entrusted François to the Jourdan family, to Camille, his wife Louise, and their 18-year-old daughter Yvonne. Painful as it was to contemplate separating from their youngest in such threatening times, they knew it would give their son the best chance to escape the wrath of the Nazis. They left with Marc for another hiding place, not far away, at an elderly Flemish couple's house, the Van Cauwenbergs, who were already sheltering another Jewish family. To François, they explained in few words that they were going back to Brussels and that, for his own security, there should be no contact between them. Prior to their departure from the Café-Restaurant, François was already instructed to pretend not to know them. Despite the sketchy details provided by his parents, François was fully aware that his family was running a huge risk. He understood perfectly his clandestine situation: officially, he was a "nephew" of the Jourdans'. But the situation remained perilous. When a customer at the Café-Restaurant asked him insistently, perhaps out of mere curiosity, his name and address, he responded spontaneously "Englert." "Oh," said the man, "the same name as those folks over there." Unfazed, François replied, "I don't understand why they would give you that name!" And for his address, he improvised a street name he must have heard at some point, since that street really does exist in Brussels.

For an entire year, François did not attend school, but upon his parents' insistence, he took private lessons at the home of a former schoolteacher who lived right next to the Café-Restaurant — he could go back and forth without risk. He would have a few other rare occasions to move about outside the Café Restaurant, notably to tend to the Jourdans' sheep. Surprising for a young city-dweller, François loved animals and knew how to communicate with them.

Yvonne taught François the rudiments of piano, which he would continue to play all his life. François understood that any contact with the outside world came with grave risks. Torn from the family cocoon, he was miserable and anxious. He missed his parents, their protection. The

Jourdans were kind to him, but he still felt alone and vulnerable. He had nightmares, started wetting his bed again, and overate to calm his anxiety. The few incidents he experienced in Brussels came back to his mind, as was his emerging awareness of belonging to something peculiar that aroused hatred, something that needed to be hidden at all costs: the fact of being Jewish.

Unbeknownst to François, Yvonne was sending news regularly to his mother who, defying the odds, would come visit him secretly, making a detour through the train station to make it appear as if she were arriving by train from Brussels! Fortunately, Yvonne and the child got along well. At her side, François even participated in some Resistance actions.

Yvonne helped the Resistance prepare the sabotage of German trains, which required someone to precisely describe when and where the trains were passing, and François was put in charge of counting the exact number of cars in each train. After the war, Yvonne was awarded a medal and a "Resistance Certificate."

Among those people willing to risk their lives to save another, we need to cite a friend of the Jourdans', a bachelor named Achille Moreels, a local news vendor, ferociously anti-clerical, a strong-spirited man known for peppering his speech with Rabelaisian expressions — François recalls an odd metaphor where anything and everything was compared to "the Pope's balls."

Achille was the only person François spent any time with apart from the Jourdan family, and a close friendship developed between this solitary man and the banished child.

A long time later, in the company of his second wife Danielle Vindal, François dropped in on him unannounced and was very moved to find his childhood photo sitting atop a chest of drawers.

In 1943, Baltche went to Brussels to finally get the family's fake IDs, made out in the name of Englebert, natives of Petigny, a small Walloon village. Joseph was made into Michel-Simon and Bajla, Marie-Madeleine. Upon her return, she was told of tip-offs and arrests — someone in the Resistance with whom they made contact had been arrested — and she decided they could no longer stay in Lustin.

The parents set off immediately to get François, in the middle of the night, and the family found refuge at Achille's. Baltche set her little Fajwel on her knees and an emotional François recovered that sensation of shelter that only family can provide. The bedwetting stopped but the nightmares continued.

Achille learnt from Madame Van Cauwenberg that the very night they fled, the Gestapo arrived at dawn to search their house!

Here began the wanderings of the refugees who became fugitives:

The family found shelter for one night at a farm over near Crupet. The farmers set up a tent in a field where their 15-year-old daughter would spend the night with François who, lulled to sleep by the chirping and twittering of the local fauna, will never forget this dazzling night under the stars.

The family then stayed with a brother of the Jourdans' in Stoumont, in the Ardennes. There, they witnessed many skirmishes between the Germans and the Resistance fighters. They could hear shots fired in the forest nearby. The Germans were in the village and the accommodation-provider was starting to panic.

It was time to press on once again.

In the village of Annevoie, the father found a house to rent. What they needed to do was blend into the scenery. The father decided to confide in the local vicar, Abbé Warnon. Suspicious at first, the priest mistook Joseph for a troublemaker. But once he understood the situation, he took the family under his wing.

The cleric had the family registered in the commune under the name Englebert, which made them eligible for the much-needed food stamps.

Without really understanding what was happening, since the parents tended to withhold explanations, François was nevertheless fully aware that they were living under a terrible threat — he felt death ever-present, haunting his nightmares, but strangely, he was unafraid of it. One morning, he woke up with a terrible bellyache, and immediately concluded that it was something deadly, which did not scare him at all. For this ten-year-old youngster, death had already assumed the allure of deliverance. He imagined how sad his parents would be, how much they would miss their child, and decided to be extra nice with them. Baltche noticed the change immediately: "Oh, the little one must be sick to be so nice!" The child was taken aback by the anxious tone of his mother's voice, an anxiousness that would remain a part of their lives from then on.

But they had to stay the course.

Life continued in Annevoie. Charles Krivine brought them a radio. Abbé Warnon stopped by regularly to listen to the news programs in French on the BBC and Radio Moscow. Victories by the Red Army were a source of hope for the Resistance fighters, who included many Communists and Jews (Communist or not). For Abbé Warnon as well, his hatred of the Nazi occupiers was such that, although he had no sympathy for the Soviet Union, he rejoiced in its victories over the common enemy.

Abbé Warnon became a family friend.

He enjoyed Baltche's Yiddish-spiced French.

When he quoted the proverb, "You have to aim high to not fall back too low," Baltche transposed it into Yiddish for the benefit of the family: *Men darf pishen hoyer, falt es nit tsu niderik ...*[25] But beneath his *bon vivant* exterior, lover of good food and fine wine, hid a tormented soul. Rumor had it that he entered the priesthood after an unhappy love affair. His sincere faith could not obscure a certain bitterness, which he expressed in the bluntest of terms: *If God doesn't exist, then I've been screwed over royally!*

His sermons were fiercely intransigent, and one can assume that when he encountered little François, beyond the courageously humanist act of saving a child in danger, there loomed the profound desire to save a soul.

In the Annevoie house, life took on an almost "normal" air; they had a vegetable garden, they dared to venture a few steps outside, and there was even a little cat that followed François everywhere after he saved it from drowning, and which performed little acrobatic tricks that François taught it.

But legal obligations remained in force even in occupied Belgium.

---

[25]You have to piss high, for fear that it will fall back down on you.

In autumn 1943, shielded by his new identity, François had to go back to school.

Abbé Warnon interceded on his behalf with the school's principal to get François enrolled in 7th year elementary at the Collège Notre-Dame de Bellevue in Dinant, an imposing brick building perched on a hill overlooking the little town, stretching along the banks of the Meuse.

His young age required that he be a boarder. Torn away once again from his family, he was subjected to the daily horrors of boarding school, where the day began at 5:50 am with summary ablutions, clothes thrown on, study hall at 6:00, followed by Mass and another study hall, and finally at 9:00, breakfast. Talking was prohibited during lunch until the end of the sermon; talking in hallways was also prohibited — in short, talking was not allowed practically anywhere at any time.

The piano teacher, a layman, would rap the knuckles of pupils with a ruler. Pupils were slapped with a vengeance. It was so cold that they would sleep in their socks — which was also a sin if one were to believe the stern spiel delivered daily at breakfast.

The change of scene was complete; the child was completely disoriented. Once again, he found himself plunged into a world of solitude, beset by bewilderment by day and the heartache of abandonment by night. Most especially, he failed to understand the purpose of this cruel discipline, with no logical connection to the clerics' soothing message of "divine love, love thy neighbor, love, love …" He would harbor a deep distaste for the icy atmosphere and hypocrisy of institutions of this sort, an abiding distrust of collective living and an abhorrence for the word "group."

And then, there were the duties inherent to life at a Catholic boarding school: Mass, communion, confession …

In order to take communion, one had to be baptized. To allow François to live at the school without committing, in Abbé Warnon's eyes, a sacrilege, the cleric went to Namur to get the bishop to grant permission for François's baptism. The bishop, Monsignor Charrue, a member of the Resistance who, to his great credit, excommunicated Léon Degrelle,[26] of course granted the request.

---

[26] Léon Degrelle, founder of the Rex movement, active in Belgium from 1935 to 1940. A nationalist party close to Catholic circles at first, Rex became a Fascist party during the Second World War, grew closer to the Nazis and was soon collaborating openly with the occupier.

Thus, the child started praying to God, and for this child who was feeling so alone, so far from hearth and home, prayer became a gentle companion, someone he could talk to and express himself, and it did him a world of good. He sang in his school's chorale, and happily so, being a great music-lover. In such rare moments of grace, he felt he had a "religious soul." A few shafts of light to brighten his daily ordeal.

Other duties proved more problematic.

At confession, he was supposed to own up to his sins; to help the children, they were furnished with a booklet of all possible sins! François was puzzled. He didn't find any sin that applied. After much hesitation, he finally came up with one, and told the priest, "I touched what I was not supposed to touch," without really knowing what that could possibly imply. The priest leant toward the child, his nose pressed onto the grill-work, to explain that at times, of course, one is obliged to touch it for certain needs, for example, but that this is not a sin, provided you don't experience any pleasure.

The boarders rarely left their residence, except for major holidays, summer vacation, and certain weekends. His brother Marc was more fortunate: as he was much older, he was able to enroll as a day student, had a room in Dinant, and biked back to Annevoie on weekends. He also sometimes biked over to François's school to pick him up, for François had his own bike, too! The parents had one made for him in the village, cobbled together with odds and ends by a neighbor.

When the two brothers were back together with the family, biking was their favorite thing to do. These bike outings provided the space they needed to feel free, but even then, they could never let down their guard, for the village café was run by a Rexist.[27] The brothers had to be careful to take only the backcountry roads to avoid passing in front of the accursed café. And even the bike itself was not to be trusted, as they found out once when, racing madly, a wheel detached and sent François flying. It was a hard landing, and François passed out. Underneath his bushy eyebrows, the scars of stitches — done by a veterinarian at home, since going to a hospital was out of the question! — are still visible to this day.

---

[27] Partisan of the Rex movement founded by Léon Degrelle.

In the little Annevoie house, François's absence weighed heavily on the parents, who felt the need to see their youngest more often. One Sunday, they risked their lives to go see him at his boarding school.

François was deeply unhappy there, and troubled. The dehumanizing environment was getting him down. His nightmares were recurring, always the same: Nazi uniforms, manhunts, executions. The anxiety was unending. He missed his family, his refuge home where his little cat was waiting for him, where "life was almost normal."

François had barely any contact with the other boarders, solitary and an avid reader, much as he was before the war. It was only much later that François learnt that there were other Jewish boys hidden at his boarding school. Only one incident ever arose: a boy came at him one day with "Dirty Jew, you don't belong here!" François didn't react. After much hesitation, he decided not to tell his parents about it.

In February 1943, the massive German debacle at Stalingrad made the prospect of an enemy defeat seem possible. Every night, British and American bombers flew overhead. At the Annevoie house, hopes were high but tenuous. Worry was constant.

In 1944, Marc was summoned under his false identity for the STO[28] to Germany, as were most 18-year-old Belgian males. Aware that it was risky for Jews to find themselves anywhere in Germany, Marc made up his mind not to comply.

The only way to avoid the German labor camps was to join the seminary. Abbé Warnon stepped up once again, using every personal connection he had to get him into the Burnot Seminary in Profondeville. Marc, a total non-believer, had to abide by all the requirements as the future servant of a religion that he rightly considered one of the main causes of hatred of Jews in the Christian West. But he was also in a position to distinguish between the Church and its faithful, among whom many, like Abbé Warnon, were helping save the children of the "deicide people, object of divine wrath."[29]

---

[28] Service de Travail Obligatoire, or Mandatory Work Service, established in Belgium in October 1942. Deportation of workers to Germany from the newly occupied countries to fill the labor gap caused by the massive mobilization of Germans sent to the Eastern Front.

[29] Expression used for many centuries within Christianity to designate Jews, a use that the Second Vatican Council put an end to in 1965.

June 1944, the Landing at Normandy! Anglo-American troops poured onto the French shores.

At the Burnot Seminary, as in the little household at Annevoie, everyone was tuned feverishly to the news from the BBC.

Burnot shut down.

Hope became reality.

But the Germans hadn't left yet, and they committed numerous atrocities in the Ardennes. Chaos reigned, with Germans and Resistance snipers shooting it out. The area surrounding the Englert's house became a danger zone. The parents and François took refuge with Abbé Warnon in a house belonging to some people from Brussels, at an off-road location. There, they were joined by Marc, reuniting the family almost miraculously — as it turned out, the Burnot seminarians were evacuated to the surrounding farms! Another miraculous homecoming: the little cat and François are back together again!

One day, François heard bullets whistling past him. The routed Germans were stampeding through the region, firing on anything that moved, burning villages to the ground.

François, who by now was eleven, was witnessing a bewildering situation: from their house on the edge of the forest, they were hearing the radio announcing triumphantly the successive Allied victories, applauded by the whole world, while right outside their house, machine guns rattled and bullets whistled.

Will the war never end?

September 1944, Brussels was liberated and the Americans finally reached the Annevoie region. In December 1944, the Germans set off the dreadful Battle of the Ardennes, but for most Belgians, it was already Liberation.

The family, together again in Annevoie, was ready to head back to Brussels in October 1944.

Joy, emotion, relief, but also apprehension: what had become of their friends? As for those left behind in Poland, they dared not even imagine.

The Englerts found their apartment intact. It had been placed under seal at the start of the war and remained so throughout. As it happened, after the family made their hasty departure in July 1942, the German police showed up at Boulevard d'Ypres, noted that the premises were unoccupied, and sealed it up. The apartment was then "forgotten" by the German authorities.

The sealing order had nothing to do with the fact that they were Jewish, since Jewish assets were systematically and unscrupulously pillaged by the German occupiers; rather, it originated from the fact that the Germans, after their raid of the Café du Vieux Rampart, were apprised of the Englert family's contacts with the Resistance.

I.5. The Righteous

The Jourdans did save François, but it is obvious that the opportunity for the parents and Marc to be sheltered by going underground at the Van Cauwenbergs', along with another family, in the Lustin region, was only possible thanks to the complicity of the entire village. Getting fresh supplies for that many extra people could not have gone unnoticed, and it went without saying that food stamps were not available to the Englerts. It took many brotherly and courageous people for them to have escaped the terrible fate of so many Jews — but alas, there were the unscrupulous ones as well, for so many of those in hiding were reported and turned in. The bonds forged while under German threat went slack somehow once peace returned. After a few timid and ultimately fruitless attempts at reconnecting with their rescuers during the post-war period, François gave up, undoubtedly beset by the unconscious but compelling need to occlude, a need that most victims of Nazi barbarity experience. Still, time did its work:

> *As I've grown older, I have found inner peace, and am more open to others. I've begun feeling less vulnerable whenever the issue of the Nazi years is raised. The memory of those magnificent people, to whom I owe my survival and that of my parents and brother, has started to haunt me. I regret not having thanked and honored them more, or passed this memory down to my own children.*[30]

---

[30] Speech before the House of Representatives, 23 January 2019 (I.2).

In 2013, thanks to his being awarded the Nobel Prize in Physics, which brought media attention to his name and story, Claudine, daughter of Yvonne Jourdan, who had been attempting to piece together her mother's past, wondered whether this François, recipient of the Nobel Prize, might be the little boy her grandparents had hidden. Through her contacts, she came across a mutual friend thanks to whom she was able to track down François. Their meeting was more than a surprise; it was a very emotional resurgence of the past. And Claudine would become a close friend to François and his wife Mira.

Likewise, it was only much later that François would reconnect with the memory of the courageous Abbé Warnon. In 2012, Mira chanced to meet with Professor Dominique Lambert, native of the same region as the Abbé, physicist and philosopher of science at Notre-Dame-de-la-Paix University in Namur. Through him, François would learn that Abbé Warnon was very active in the Resistance. He compiled relevant documents and photos in order to constitute a dossier at Yad Vashem[31] and to have Abbé Warnon acknowledged as one of the Righteous among Nations. He would take the same measures for the Jourdan family.

As for Achille, a bachelor with no known family, who died practically without a trace, compiling a dossier would prove problematic. But Yad Vashem took the lead and saw to it that his name was added to that of the Jourdans and Abbé Warnon.

---

[31] The World Holocaust Remembrance Center, located in the forest of Jerusalem. One of its missions is to honor "the Righteous among Nations," i.e., the non-Jews who helped save Jews whose lives were in danger during World War Two.

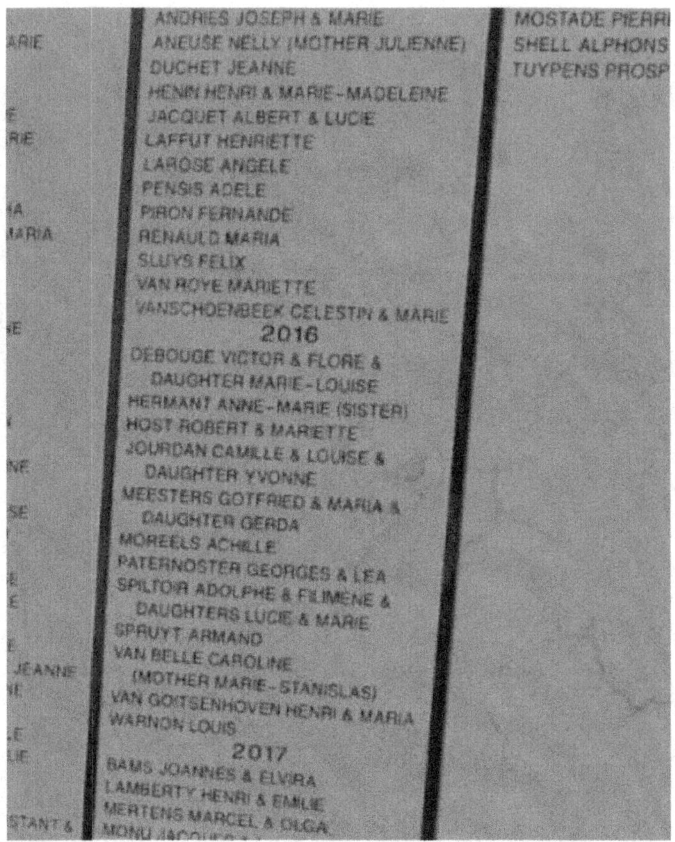

The Belgian Wall of the Righteous at Yad Vashem.

Intermezzo 2: A Memory Riddled with Holes

As François is telling me his story, I am trying to imagine the child he once was. The unhappy child who ate too much, living a vulnerable underground life, feeling abandoned despite the best efforts of the Jourdans. The little lad who refused fear, but whose recurring nightmares gave evidence of a deep distress.

The war was over, but the end brought little relief. Just the opposite. The world was discovering the scope of the genocide. In Europe, despite denials, the Nazi's resolve to annihilate the Jews was a known fact, and those in high positions were fully aware of its implementation, and some were willing accomplices. In fact, anyone with any political education at all must have suspected that something was afoot, as did most probably the people living in the villages nearest the camps. A few brave people that today would be called "whistleblowers" did their job, fiercely and at times risking their own lives — in vain, alas. In Europe as elsewhere, a combination of political reckoning, indifference, and fear had the effect of "dismissing" the whole genocide question. For the "mainstream," the liberation of the death camps revealed a reality so heinous as to surpass even history's most notoriously criminal acts. The Jews who succeeded in escaping the worst found themselves confronting the absolute horror of the Shoah. Most would dwell for a long while in an unbearable state of uncertainty with regard to the fate of their loved ones. It would take weeks, sometimes months and years, before survivors of the camps found — or not — the members of their families who survived. And when such unlikely reunions did take place, they were unable to find the words to describe what they experienced. A story too hard to tell? Most certainly. But especially too hard to hear. Need we recall that the survivors were not always met with empathy? No one had the desire or the stomach to listen to them tell their unbearable stories. Let the unspeakable remain unspoken, now and forever!

The survivors' experiences were hurriedly cloaked in silence. And not only the experience of the camp survivors but also that of all the other victims as well, and among them, the hidden adults and children — despite the extraordinary privilege of having evaded the camps. But nothing is ever completely erased, and all those concerned would have felt memory gnawing at their soul — a memory riddled with holes where the icy winds of angst come rushing in.

The Shoah also signified the destruction of a continent stretching from Alsace to Ukraine, from the Black Sea to the Baltic: Yiddishland, that vast archipelago where everyone spoke Yiddish, vector of a lively culture during the interwar period. Thousands of villages and neighborhoods wrecked and plundered, emptied of their population, where their language would never again resonate. No imaginable return for the survivors. A deep sense of uprootedness would freeze their words in time.

Thanks to documents salvaged, miraculously, from the disaster, thanks to recovered photos, we can today "see" this vanished continent. In black, white, grey, or sepia, the colors have also disappeared.

At the Englerts', all memory of the war seemed "erased." What had become of the family left behind in Poland was a subject that no one ever brought up. The parents knew: not a single family member survived. But how could they have grasped the inconceivable? Unanswered questions haunted them; throbbing unabated, the thought of this absolute catastrophe persisted as an undercurrent, the total destruction of their world, wounds that Baltche and Joseph buried in the deepest corner of their hearts. And François spoke to no one of his recurring nightmares, peopled by ghosts in SS uniforms pursuing and slaughtering, most probably the cause of his adult insomnia. But this world of a past riddled with holes, this world of "deaf-mutes" instilled in him an attitude of defiance toward the adult world and an overwhelming anxiety that he never shook off. As he himself would say much later on,

I could live an apparent normalcy only by erasing this past in my conscious life. Even the reference to those extraordinary people who had saved my life could never wipe away the imprint of Nazism's abject violence. What alone lingered in the remains of my stolen childhood was a nebulous anxiety mixed with aggressiveness, the tendency to contest anything that the world of adults wished to impose on me and the vague and ultimately futile desire to never have anything to do with it.[32]

I.6. A Sense of Apparent Normalcy ...

In October 1944, after a tumultuous five years of schooling, François was able to enroll in 6th year Latin at the Athénée royal de Koekelberg. In January 1945, the ULB reopened, and Marc could start his first year at medical school.

Life resumed its course, at least in appearance. Parents at work, children at school. Despite their unshared mourning, Baltche and Joseph did not abandon hope to hand down to their children the traditions of the culture into which they were born. But François and Marc rejected outright the traditions linked to Judaism. As for religious rites, although Marc agreed to do his bar mitzvah[33] without too much fuss

---

[32] Speech before the House of Representatives, 23 January 2019 (I.2 and I.5).

[33] Rite of passage for Jewish boys, roughly equivalent to Confirmation in Christianity.

before the war, this was not the case for François. In preparation for the ceremony, he was sent to take special classes, where he got expelled two days later. A private tutor was hired to teach him the rudiments of Hebrew required to be able to read the Torah — to no avail. The tutor gave up after a few days.

The parents were disappointed, but hardly surprised. Everyone in the family knew that François did not yield easily when it came to imposing anything on him that was not to his liking.

Already, before the war, it was clear to everyone that François would never be a "good lad." Though he dreaded being alone, he liked to figure things out for himself and saw no need to obey anyone at all. The anxiety and defiance that the war instilled in him only exacerbated these character traits.

Marc was more accommodating. Even while he was at medical school, he listened patiently to his father teach him the basics of the *shmattes*[34] trade, since one never knew what the future might hold …

But however much François revolted when his parents attempted to pass on a few bits of secular Judaism mixed with nostalgia for the traditions they brought with them from Poland, he would never deny his humble roots. In fact, he felt comfortable in his family milieu. Later, he would feel no particular admiration for intellectuals, no yearning to join the "cultured Brussels bourgeoisie," even though the cultural capital he had accumulated would have invited him to do so. He would always feel infinitely more at ease among craftsmen, tradespeople, technicians, artists, marginal and emancipated types in milieus where you were free to be yourself.

I.7. A Not So Studious Adolescence

Because François had become Catholic during the war, he was enrolled in a class on religion at the Koekelberg Royal Athenaeum. Was this a personal choice, or was it Baltche's decision, a gesture of gratitude toward people like Abbé Warnon? Whatever the case, two years later, François

---

[34]Yiddish word meaning fabrics, rags, and clothing.

decided on his own to drop religion in favor of a class on ethics. The same year, he opted for the Latin-Math stream.

At the Athenaeum, François discovered the spirit of camaraderie that prevailed among children of the same age, a novelty for a boy who experienced only the solitude and anxiety of the war years. Those years of living dangerously, now erased from his conscious life, were unforgotten nevertheless, for every night, his nightmares plunged him once more into the horror. Never again would the peaceful nights of his pre-war childhood return. His life was set at two different tempos: the chaotic rhythms of the night and the imperative daytime tempo — because, during the day, he had to live "as if nothing were the matter." This was his personal form of resilience, the way he managed rather well, in spite of it all, to fit into his school and make friends with a few classmates, but without ever going so far as to raise the issue with anyone about his past as a hidden child.

As for the education he received, he has no fond memories.

Even his math class held little interest, except as an upperclassman when he had the good fortune of having an excellent teacher. His Latin class was rowdy, and François gladly took part in the rowdiness. He found certain classes downright ludicrous, such as his drafting class during which the students organized burping contests. He would later come across his indulgent drafting teacher, sitting at a table in a bistro, sharing a mug of beer with his dog.

He didn't study much.

However, he did act in theatrical productions put on at the Athenaeum; the female roles were performed by students from the Lycée Royal de Molenbeek — at that time, coeducation did not exist, and every time the high school girls entered the Athenaeum, bewilderment ensued. Molière's "The Imaginary Invalid" was a huge success at the school. Among the young male actors was Jacques Pivin, future mayor of Koekelberg, and among the girls was Rose Leszinska, who would remain a lifelong friend.

But most especially, he read ravenously, and discovered a passion for poetry: Baudelaire, Verlaine, Eluard, Aragon, Apollinaire, learning dozens of poems by heart.

Baudelaire would accompany him throughout his life. *Les Fleurs du Mal* constantly springs to mind, and he quotes it at length. One day, a first line emerges in the midst of one of our conversations ... the memory hesitates an instant, then the entire poem resurfaces, layer by layer. François smiles, his eyes twinkling. He experiences an exquisite pleasure in recovering familiar prosody.

Family vacations were spent at Knokke, a sea-side resort. Baltche enjoyed gambling at the casino; she even sometimes bet François's allowance!

But François hated vacationing at Knokke. Here again, he sought refuge in books. At the beach, his reading list was quite serious, Marx, Engels, and Lenin. Marx's critique of capitalism appealed to him intellectually; he wanted the world to function according to Marxist principles, but was less enthusiastic about the theory of use and exchange values, very flawed in his view, and thought it was unfortunate that Marx glossed over the psychological aspect of social phenomena. For François, Marxism was attractive, though ultimately unconvincing, and he would be less and less convinced as the years passed. Too much disillusionment tarnished the image. Still, whenever François is challenged by some critical event, he would reach into his heart for the same ideals of humanism and solidarity that he so long hoped to find in leftist ideologies.

Despite his school chums, the lonely child grew into a lonely teenager. The presence of a brother did not make much difference, since practically, they only ever talked about music. Music and reading were François's two main sources of spiritual nourishment. He continued playing the piano with great joy, and his parents, always concerned about their children's cultural development, found a second-hand piano, an Erard, and hired a private music teacher. François practiced two hours a day. A kind lady, Madame Delecluse, brought him scores. She was the wife of the family's only non-Jewish friend, to whom the father served multiple little glasses of spirits, since in his view, this was how one entertained a "goy."

This old piano, a beautifully crafted cross strung instrument, would follow François his whole life. It was his most precious childhood souvenir. He has never been able to bring himself to part with it, despite its time-worn state. Later, when he finally procured an excellent Kemble, he had one of his children take the Erard, and after that, it went to the ex-husband of his youngest daughter. Then, one of his grandchildren practiced his first scales on it. The piano still exists today, but they did sell it in the end. Years later, upon learning this, Marina Solvay,[35] who had become a friend, would regret the disappearance of this piano which, in addition to its historical value, was "the first piano of our first Nobel Prize in Physics."

The adolescent, who so closely observed the world of adults, perceived them as each having a fixed, well-defined "personality." One day, while he was walking along the Boulevard Botanique in Brussels, he vowed that no one would ever confine him to this thing that society seemed to hold in highest esteem: a "personality!" François, only fourteen years old at the time, made this momentous foundational decision: it would be up to him, and to him alone, to construct himself. Or not even really "constructing," but rather moving forward according to his inspiration, gleaning wherever he pleased the knowledge to which he aspired, soaking up books, music, school lessons, in search of answers to his wide-ranging questions.

---

[35] Marina is the great-great-granddaughter of Ernest Solvay, a Belgian chemist and philanthropist, founder of the industrial group Solvay and Co. and creator of a series of conferences in physics, known as the Solvay Conferences, which I will refer to later.

Episode II

François and the Others

II.1. Abhorrence of the Polytechnique ... *et la maison du bon dieu*

Encouraged by his math teacher at the Koekelberg Athenaeum, Monsieur Heuchamps, and by his parents, François went into engineering at the ULB in 1950. Monsieur Heuchamps had organized extra math lessons for students wishing to sit for the Polytechnique entrance exam. Most of the Latin-Math students prepared for it and all were admitted with ease.

At the Polytechnique, as at the Athenaeum, François was hardly the model student.

He regularly skipped class, but managed to play his cards right at exam time. In his third year, for instance, he showed up for his Paleontology exam knowing nothing of the material. The professor presented him with a document to comment. François had nothing to say. The irate professor finally threw a matchbox at him, hitting him in the head. Suddenly, inspired by the memory of something he had read in a Soviet textbook, François started talking about the theory of evolution and Lyssenko's infamous blunders regarding his theory of the inheritance of acquired characteristics. The professor was duly impressed and François passed the exam with flying colors.

In his fourth year at Polytechnique, neither François nor Jean Finn, a friend from back at the Athenaeum, had taken any notes in their materials science class. Fortunately, Jean Finn had a very devoted father. The night before the exam, the two accomplices managed to somehow get hold of a detailed syllabus, which they studied in pieces, each at his own home, with the father shuttling back and forth to bring the already studied bit to the other, and vice versa. By the end of the night, François and Jean were ready for the exam … and the father was completely exhausted!

It was at the Polytechnique that François met another future physicist, Joseph Katz. When François was just starting, Joseph was already in his second year, but due to an argument with a professor whose class he refused to attend because he found it incoherent, he was failed and ended up in the same year as François. They didn't become friends right away — Joseph hadn't fully processed his failure and was constantly grumbling — but they had much in common, notably their anti-conformist streak, their contempt for academic (and dress) codes, and their revolt against the brutality of the capitalist world. François would later learn that Joseph had also been a "hidden child."

The Polytechnique, its arrogance, its tedious atmosphere, its drinking parties that followed a strict academic agenda, its teaching obsessed with the technical application — not always the most rigorous — of theoretical knowledge, none of that was at all to their liking. Neither François nor Joseph was even remotely involved with conventional student life.

Fortunately for François, he had great opportunities to get away from all that.

Just as his dislike of family vacations had hurled François into the arms of Marxism, his equally strong aversion to the Polytechnique drew him to a group of dissenters of another caliber, where François would discover his true fellow travelers.

Those were the glory days of the youth hostel movement — left-leaning or even anarchist young people were enthusiastic followers. François and Jean Finn, along with other students who enjoyed long bike rides with friends, had been patronizing them for quite a while. At one of them, François met a beautiful and intelligent woman, Nina, and her husband Robert De Mayer, a cabinetmaker specializing in fine "antique" furniture. Nina was Jewish, Robert's parents had hidden her during the war, and Robert married her upon his return from the STO labor camp (I.4), where he had volunteered so as to avoid the authorities seeking him out and risking their discovery of Nina. This encounter would lead François to similar ones that would have a decisive influence on his life.

Robert De Mayer was a warm, generous man, and very uninhibited when it came to all things sexual; he spoke openly of his problems, which he blamed mainly on his wife, and was searching feverishly for a "lifesaving nymph." Nina, Robert, and their two children, Nicole and Michou, shared their house with Robert's mother Mémé and Victor, his schizophrenic brother, who helped to "distress" the furniture. Nina was very fond of Mémé. Mémé, for her part, took care of her ailing son whom she refused to institutionalize despite serious episodes of delirium. Mémé would commit suicide at ninety after Victor's heart attack left him for dead. They found a note where she explained that she had lost her reason to live.

Robert and Nina's place was always open, *la maison du bon dieu*, as the French say, *de zoete inval*, as the Flemish say, a very welcoming haven. People from everywhere showed up there. Together, Robert and Nina formed the core of a group of friends that would develop over all the years that François spent at the ULB, and well beyond. A little tribe of sorts, bringing together artists, craftsmen, disenchanted academics, marginals of all stripes, a motley crew of creative, flamboyant people, most of whom demonstrated a definite penchant for anarchy. These were the ones François socialized with during his college years, and who helped him survive the ambient grimness of the Polytechnique. Over the years, this second family deployed like a wave of concentric circles, enriched by countless friends coming from all fields and from all corners of the globe.

There were many Jews in the little tribe, François noticed. Why so many Jews among the marginal, the avant-garde artists, the rebel intellectuals? He was compelled at this point to ask some questions about his own Jewishness, without being overly preoccupied by the issue.

For these new friends were bringing him something totally other: Georges Miedzianagora the philosopher and his partner Jacqueline; Annette and Yvon, she a painter and he a doctor; Jerome, another philosopher; the physicist Harry F.; Joseph Katz and his wife Anne-Marie; the beautiful Anita and Jean-Jacques the musician; the writer and human rights advocate Marek Halter and his wife Clara; the painter Arie Mandelbaum; Cornel the Hungarian, huge poker player, and his partner-to-be Diane; the attorney Roger Lallemand; and the owner of the restaurant La Maison du Cygne, right on the Grand'Place in Brussels, a larger-than-life fellow, genius chef, alcoholic, cabinetmaker, renowned calligrapher, painter, a rebellious soul who refused to send his children to school so as not to expose them to the "imbecility of compulsory education," and who would die of cirrhosis of the liver.

One night, recounted François, *after the restaurant had closed, he uncorked a bottle of Château d'Yquem for Georges, Esther[1] and me. We talked and talked, reinventing the world as we emptied the bottle. In the early hours, we discovered the Grand'Place completely depopulated, sound asleep, a delightfully cool breeze wafting through the empty space — an unforgettable memory.*

There was also the attorney N. who loved to head out on Saturday nights to explore the seedier side of Brussels, sleazy cafés patronized by hookers, delinquents and mafia types, dismal asylums that provided refuge for the world's misery, where vagrants slept "on a rope" as in the 19th century — at 5am, the proprietor unhooked the rope and the sleepers woke up on the floor.

Close association with so many atypical temperaments meant that François's anti-establishment tendencies and pensive turn of mind could flourish in total freedom. He also discovered the pleasure of sharing. He was finally shedding his life of solitude, and his anxiety was loosening its grip.

II.1.1. *Georges Miedzianagora*

A central figure of the circle was philosopher Georges Miedzianagora, one of the assistants (along with Pierre Verstraeten) to Professor Chaim Perelman, a fascinating, provocative figure. During the war, when he was only 10, he and his mother managed to escape the Malines transit camp located in the former Caserne Dossin barracks. He never spoke of his father. The war had deeply traumatized his family. Georges, despite his brilliant mind, sharp wit, and sense of humor, was a wounded being. He had moments of deep distress and panic attacks. After the war, a high-achieving student, he studied law and philosophy at the ULB, got married, left for Israel with his wife, and then returned to Belgium after a short stay

---

[1] François's future first wife.

on a kibbutz. His wife soon left him and he started living as a couple with Jacqueline, a lawyer.

Georges, who could be terribly aggressive in his speech, got into fights with everyone. Robert the cabinetmaker, for instance, although he did value Georges, sometimes felt uncomfortable around this erudite philosopher, who was at times sarcastic, or even nasty, and at other times warm and caring. But between Georges and François, a deep friendship would emerge. There would never be a hint of conflict between them.

It was through Georges Miedzianagora that François understood that it was possible to learn from others. Until then, grappling with his anxieties, he had remained to a great extent in his own world, often bewildered by those around him.

He was impressed by the way George had of understanding people, getting close to them, figuring out what made them tick. Georges opened up for François the world of others. They engaged in long philosophical discussions, and even co-wrote an article that would appear in *Les Temps Modernes*: "Sign, sense, dialectic."[2] In philosopher circles, the article earned praise for its subtlety but became a subject of controversy at the periodical.

Georges's friendship allowed François to flourish, far from the constraints society was attempting to impose. The two of them reserved the right to decipher, critique, assess, and contest the world around them. Though anarchist ideology did appeal to them, they were not interested in belonging to a movement or party: the exercise of total intellectual freedom was paramount.

But François's spirit of dissent was no idle theory; he also put it into practice.

In 4th year Polytechnique, he did an internship in England at a large cable manufacturing plant. His job consisted of checking and calibrating the machinery, but he ended up servicing these machines, which were in appalling condition.

His father, proud of his Polytechnique son, wrote him letters stuffed with advice that shed light on François's stay in England in 1953:

---

[2] *Les Temps Modernes*, no. 217, June 1964.

Dear François,

"We got your letter which we were very happy to receive, seeing that you're having such a good time, but upon reflection, we get the impression that you are not there for an internship but for a vacation, since you haven't said a word about what you are doing at the plant. Are you learning anything?

...

...Be careful not to catch cold, and don't go to bed too late ...

[...] and don't do anything reckless at those meetings in Hyde Park, it could lead to problems...

Sending all our love, as are Marc and his family,
Your parents"

François was discovering the abject poverty of the English working class, the legendary Satanic mills, the noise, the filth, assembly-line work ... and the sacrosanct tea time.

He found the workers' situation untenable, and felt their trade unions lacked a combative spirit. Along with an Iranian colleague, they wrote and distributed tracts, and succeeded in mobilizing the workers and organizing a strike that was followed massively by the workers. François was summoned by the management, who were furious. They proposed that he leave immediately, with the agreed upon remuneration. Nothing could have made him happier. He was overjoyed to get back to his friends in Brussels.

His friends who were having a pretty wild time occasionally. They were making wacky 8-mm films, a mixture of schoolboy prankish fun and surrealist burlesque. They played wild poker games with extravagant bets. Jerome the philosopher, obsessed by the idea of creating a "real left-wing journal" which would never see the light of day, paid his gambling debts with bouncing checks. Harry F. the physicist, ruined by his gambling habit, would end up exiled in Israel by his father, where, more or less settled down, he would become a professor at Bar-Ilan University, a school that strove to reconcile the Torah and universal knowledge, where theology and Judaic ethics rubbed shoulders with the most up-to-date technology classes.

II.1.2. *Paul Glansdorff*

François introduced his friends to Paul Glansdorff, professor of Thermodynamics at the ULB, someone who would matter greatly in his life.

Paul Glansdorff took great pleasure in the company of intelligent, anti-establishment people. Beneath a rather classic exterior, elegant and easygoing, there resided a soul that welcomed the most unorthodox ideas.

The anarchist attitudes of the little tribe did not deter him in the slightest.

Later, during their "poker" parties, when the friends sometimes drifted into weirder games, Glansdorff did not shirk his role of benign witness.

Among their games of choice was a role-playing game inspired by a Roger Vailland novel, *The Law*, whose title referred to a game very popular in southern Italy. It was a cruel game, an Italian variant on sadomasochistic macho harassment, and implicitly, the ruthless struggle for power among the castes of an underprivileged society. In the role of *Padrone*, assigned at random, who will enjoy full power for the duration of the game, was none other than Georges Medzianagora, with François as his assistant.

It is at Glansdorff's home in Brussels that some of the most memorable gaming sessions took place.

Paul Glansdorff took a great liking to François, who in turn felt warmly welcomed. Whenever he was going through a difficult time, Paul Glansdorff was there to provide comfort. Their get-togethers became rituals: every Thursday, François was invited to have lunch with Paul and his family. Glansdorff also had him come to stay at their country house in Baisy-Thy.

The Glansdorffs had two daughters, Nicole and Elsa, and one son, Jean-Pierre. Nicole was a warm, free-spirited girl, an artist who exceled at coming up with funny turns of phrase. Elsa would become a lawyer. Paul's wife Suzanne, a cousin of the poet Alain Bosquet de Thoran, himself the grandson of Corneille Bosquet de Thoran, former director of the Théâtre de la Monnaie, exuded irresistible charm that the years never dimmed, a charm to which brother Marc was not indifferent …

Generous to a fault, Paul Glansdorff was always ready to assist people he liked and admired. He would later help Joseph Katz get an academic job, that top-level physicist who had a knack for getting on everyone's wrong side. He would also help untangle Georges Miedzianagora's run-ins with academic administrators after May 68. He even prefaced one of his books.

L'école buissonnière: A Big Shake-up

The little tribe that gravitated around Robert De Mayer was quite naturally moving toward a very liberated notion of love — we're talking about the early fifties here, with all their weal and woe. Alongside the merry libertinage of some, there developed a close look at the place of sexuality in life, the issue of total freedom in a marriage, and the very meaning of couples.

Couples were coming together and coming undone at a steady pace: the Georges–Jacqueline marriage was on the rocks; Jacqueline left for Africa with Cornel who fell in love with Diane; later, Jacqueline would go live with N., the attorney. Their story would end in tragedy.

Georges fell in love with the painter Annette Deletaille; he put together "The Queen Annette Notebooks" for her, a collection of playful drawings where all the various friends were included. Before his dates with Annette, he sometimes asked François for advice:

"To seduce Annette, what music would you choose?"

"Mozart, without question!" François replied.

This constantly ebullient state created a fertile breeding ground for some, but proved harmful for others who finally sank into utter helplessness. Sexual liberation did not seem to be providing men or women with the happiness they expected.

Joseph Katz was well aware of this situation; with his wife Anne-Marie, they made a very classic couple, almost formal — they even used *vous* when addressing one another, the very image of the ideal couple. Fearing he would lose his spouse among this group of free-love libertarians, he distanced himself from them and formed a new circle. Still, Anne-Marie left him in the end. Utterly traumatized, he declared to Nina:

"If that's how it is, then I'm going to live the crazy life." No sooner said than done!

What about François in all this?

François was not a fan of this "sexual liberation" he felt it heavily weighted, despite all the egalitarian lip service, toward an unhealthy dose of machismo, or even a certain contempt for women.

François, whose childhood had taught him that women were important people without whom the world would not spin as it should, he who, until he finally graduated and to his great dismay, hardly ever interacted with female company (not a single female student in Polytechnique!), this same François, despite the emotional upheaval that his inclusion in Robert's group represented, was never anything but kindly curious and respectful toward women. In his eyes, they were suffused with an enchanting mystery, and he would always value the close friendships he has been able to have with women.

II.2. A Busy Year (1955)

In June 1955, François graduated with his engineering degree, Summa Cum Laude. His final project, done together with Christian Jauquet, involved the theoretical and experimental study of antennae, and proposed an original method for broadening the bandwidth of a dipole antenna. Their work would be published in the *Revue H.F.* (1955): "Theory of the Folded Dipole Antenna." François's first publication!

With a huge sigh of relief, François wrapped up his Polytechnique years. He felt that even his physics courses had fallen short. He realized that what really interested him was understanding the fundamental causes of physical phenomena, and not dealing with their industrial applications. This conviction took shape very early and would hold true from that time forward. His life would be devoted to fundamental research, but his natural curiosity would lead him to take interest in all aspects of physics, and he understood that focusing on this or that specialized area could only frustrate him. François made a conscious career move that was both linear and multipronged, linear for the goal, multipronged for the means of achieving it.

He decided to do a quick degree in Physics. But not wanting to abuse his parents' generosity, he hoped to soon find a job. Thanks to his reputation as an excellent student, he was soon hired as an assistant at the Polytechnique to Professor Frans van den Dungen.

Although François rarely shared with his parents the twists and turns of his college life, they were well aware of his achievements and could not be more proud of his academic progress. One son a doctor, the other a Polytechnique graduate, their wildest emigrant dreams had come true. To celebrate, Uncle Leon invited François to Le Grand Veneur, in Keerbergen. Since he knew that François had his qualifying exams in physics to prepare for, he booked him a room so that he could study in comfort with no disturbance, which didn't prevent Uncle Leon from getting him to participate in little hoaxes, which he adored. For instance, he would whisper to the young and naïve upper-class Flemish girls, gesticulating as he went that "Jews have noses like this! And cloven feet, too!" With François's complicity, this excellent cyclist offered the hotel's indolent clientele the spectacle of a fake cycling lesson full of silly pratfalls, which culminated in an acrobatic finale, to the astonishment of the public. François took part in these hijinks with great relish, a way of taunting the bourgeoisie by mystifying them, making fun of them but having fun in the process.

In October, François passed his special exams and was awarded his *candidat* degree[3] in Physics.

II.2.1. *One's own family*

During the academic year 1952–1953, François became friends with Esther Dujardin, a student in Chemistry, daughter of a couple who were his parents' bridge partners. He introduced Esther to his friends. Esther and François went on a vacation together the following year and got married in late 1955. As François was staunchly opposed to having a rabbi in attendance, the wedding was celebrated by Uncle Sroultche, who enjoyed citing the grand precepts of "Jewish morality" and officiated as a rabbi. François's father made an eloquent speech before presenting the young couple with their wedding rings, "circles of eternal happiness" that neither would actually ever wear — they were uninterested in bearing signs of any sort of belonging.

---

[3] Roughly equivalent to a BSc.

The young couple would go on to live in Auderghem, first at Avenue des Citrinelles, then at Avenue Gustave Demey, a second-level apartment.

In 1956, their first daughter was born, Michèle.

The young couple didn't last long at the Avenue Demey apartment. The landlord, who lived right above them, was an insufferable grump who complained constantly about his tenant's behavior. He went so far as to require that they wear soft slippers, and forbade anyone from flushing the toilet after 9pm!

Their friend the attorney intervened to reason with him, but to no avail.

François, Esther, and their friends, among whom were Robert De Mayer, Jean Finn, Georges Miedzianagora, and Jacqueline, decided to hold a particularly unruly house "unwarming" party before the couple moved out for good.

The point was to make as much noise as possible: deafening stereo, non-stop toilet flushing.

Alcohol was flowing, everyone was dancing and singing.

The irate landlord called the police.

François went down to open the door for the police, Jacqueline went along, while the other guests sang a rousing chorus of *J'emmerde les gendarmes et la maréchaussée ...*[4] François made an attempt at mediation, while Jacqueline, though a little tipsy, put her lawyerly skills to work. She informed the police in no uncertain terms that they were forbidden from entering a house before five in the morning, the law is the law. She added with a smile: "It's just barely midnight and it's very cold outside. Too bad, you can't come in."

The next day, the police issued a complaint. In his statement, the officer declared notably, "We were still 500 meters from the house when we heard earsplitting music, and people shouting 'Flush the toilet, flush the toilet!'"

François received a summons from the police. Regarding the officer's report, he insisted on specifying that "it was Bach's Brandenburg Concerto no. 6!" But this did not impress the police commissioner.

---

[4] "to hell with the police and the gendarmerie".

François then asked for an audience with the ULB Chancellor, who immediately stepped into his role of infallible mediator. He picked up the phone, and François heard what proved to be an extremely affable conversation between the Chancellor and the Crown Prosecutor. They used the *tu* form of address, as only friends do, talked shop for a bit … and the case was closed.

Shortly thereafter, the family moved into a spacious house on Rue de Tercoigne in Auderghem. It was there that so many poker games were played among their circle of friends. During one of these boisterous events, the players would be immortalized in a painting by Annette Deletaille.

François and Annette were longtime friends, since a memorable trip to Bretagne and Normandy with his brother Marc, Annette and her husband Yvon Kenis — one of the rare trips that François took as an adult with his brother. A joyful and rewarding trip in good company: a cheerful woman, two erudite doctors, three *honnêtes hommes*, "decent gentlemen" in the Montaigne sense of the term, the perfect ingredients for initiating the ever-curious eighteen-year-old François into the wonders of medieval France. The Jumièges Abbey, the Saint-Martin-de-Boscherville Abbey and manor, Caen, Saint-Malo, Mont Saint-Michel, all those cultural landmarks would be visited and commented on in detail.

II.3. Pierre Aigrain: Genuine Physics, at Last!

In the Theoretical Mechanics unit where he was an assistant, François was assigned to study problems involving girders! Professor Van den Dungen, renowned scientist and decent fellow (he was the one who, in 1941, when he was Chancellor of the ULB, decided to shut down the university rather than bow to the diktats of the German occupier), was also a big-hearted man who understood François's endless inquisitiveness. He introduced him to the French physicist Pierre Aigrain, a pioneer in the study of semiconductors, director of the Solid State Physics Lab at the École Normale Supérieure, and who held the visiting professorship that year at the ULB. François became Pierre Aigrain's assistant. He started to learn about semiconductors[5] and wrote the syllabus of the electronics class. This was his chance to dive into statistical physics and quantum physics, two prerequisites for understanding the properties of semiconductors, but which at that time were not in the candidature curriculum. Finally, something serious he could really sink his teeth into!

By the end of the 1955–1956 academic year, he passed all his exams with flying colors without having attended a single class, owing to time constraints. Professor Goldfinger pointed out to him that he had no grade for lab. François was completely taken aback: "There were labs?" As for Professor Balasse, at first triumphant when he thought he has trapped François in electronics, then completely thrown for a loop when François explained that his equations looked a bit odd simply because they used a different notation at the Polytechnique, he had to acknowledge in the end that François's exam was flawless. The other professors never flinched, for they knew they were in the presence of an uncommon mind.

During the academic years 1956–1957 and 1957–1958, he made several trips to the École Normale Supérieure in Paris, Pierre Aigrain's home terrain, where he learnt among other things how to solder semiconductors at low temperatures, a very delicate operation.

François started co-writing with Pierre Aigrain a book about semiconductors,[6] a small 200-page book that nevertheless contained a solid introduction to quantum mechanics and statistical mechanics, where all the essential notions of modern physics were addressed. He also wrote a number of articles. Aigrain was duly impressed, especially by the original method proposed by François for studying certain electromagnetic properties of crystal, i.e., a generalization of group theory,[7] which allowed for a simple and elegant way to approach thorny mathematical problems

---

[5] Semiconductors are materials, such as silicon and germanium, whose electrical conductivity is due to impurities deliberately introduced at precise doses. These materials will revolutionize electronics and are foundational to what we know today as the field of electronics.

[6] Aigrain, P. and Englert, F. *Les semi-conducteurs*, Paris, Dunod (Coll. "Monographies Dunod"), 1958.

[7] A group is one of the basic structures of algebra. It is a set of mathematical objects governed by certain laws. The notion of "group" appears for the first time in the work of Évariste Gallois (1811–1832), a mathematical genius and fierce believer in the Republic.

that had, until then, involved fastidious calculations. This work would be his first publication as a physicist.[8]

He next studied elasticity in metals, and here again, he exhibited imagination and considerable mathematical knowhow, which allowed him to interpret the experimental results in a particularly productive manner.

Guided by an intuition that seemed to evade any rational analysis but which possessed its own kind of logic, he sensed within himself a growing predilection for simple and elegant formulations. He discovered the pleasure of playing with equations and improvising fruitful encounters between mathematical concepts and physics problems. This process would become the core of his doctoral dissertation. (It is worth mentioning that the deployment of highly complex mathematical tools is the bread and butter of modern physics. More than any other science, physics speaks the language of mathematics, as Galileo explained some 400 years ago.)

II.3.1. *Les Houches*

In 1956, Aigrain sent François to a summer program, *l'École d'été des Houches*, in the Haute-Savoie in France. This school, an offshoot of the Les Houches School of Theoretical Physics, was founded in 1951 by Cécile DeWitt-Morette, a physicist whose remarkable record began in her early days at the CNRS, France's national research center, under the direction of Irène Joliot-Curie and Frédéric Joliot.

Set up in a few isolated chalets facing Mont Blanc, and at that time quite rustic, bordering on the Spartan, something like a scout camp for budding scientists, this school attracted an impressive concentration of promising students, French and foreign, and over the years, the professors included such luminaries as Léon Van Hove, Wolfgang Pauli, Enrico Fermi, and Murray Gell-Mann. François would take classes with John Bardeen, who would go on to win the Nobel Prize in Physics that year (followed in 1972 by a second Nobel Prize!) and with Bryce DeWitt.

---

[8] In the *Bulletin de la Classe des Sciences de l'Académie Royale de Belgique*, vol. 43, 1957, "Application de la théorie des groupes au calcul du couplage spin–orbite dans les cristaux" (application of group theory in the calculation of spin-orbit coupling in crystals).

Each session aimed to review the state of world knowledge in one area of contemporary physics. The atmosphere at Les Houches exuded camaraderie and a particularly invigorating drive to emulate. There, François met young physicists who already had high-profile reputations, notably Pierre-Gilles de Gennes, future Nobel Prize winner in 1991 and whose wife Annie would later open the famous restaurant *Au Boudin Sauvage* in Orsay, France. He participated in discussions with great passion, was gaining self-confidence and starting to glimpse what path he might take to achieve the scientific objective he was seeking.

His participation at the summer school proved an important step not only in his training as a researcher but also in confirming the vocation in which he was, by this point, so invested: to understand the universe around us. He would remain forever grateful to Pierre Aigrain, his elder, who had been so compelled by François's intelligence and who gave him a leg up, in a way, and would continue to support his disciple in all his undertakings, even when they took François down an adventurous path that would lead him away momentarily from his academic work.

That being said, one day when Aigrain offered him the very well remunerated directorship of one of France's largest technology companies, François flatly refused it. Money did not interest him in the slightest, and he could not imagine for even an instant "wasting" time that might be devoted to fundamental research.

L'école buissonnière: The Tribulations of a Young Physicist in Moscow

In July 1957, François took part along with a ULB delegation in the World Festival of Youth and Students in Moscow, where 34,000 young people from 131 countries came together. Several friends of his also made the trip: Roger Lallemand and Ida, Anita and Jean-Jacques Laydu, Paul Danblon, Leon Ingber, whom they call Leontchik, and Marianne, Joseph Katz, and Anne-Marie. The train ride was endless, three days and three nights.

They slept on narrow benches.

"We should take a tip from canned sardines," declared Joseph.

He and Anne-Marie slept squeezed together, top-to-tail … like sardines, as it turned out.

Moscow!

Buses shuttled the participants to brand-new student accommodations. No sooner had they arrived, tipsy with fatigue, than the old buddies started going crazy, laughing and clowning around. Roger Lallemand started throwing things out the window, to the great displeasure of the Soviet chaperones.

The next day was the inauguration of the Festival.

François took no interest in the official speeches, nor was he dazzled by the various delegations' artistic displays. He set out on a stroll around Moscow.

He discovered, behind the palisades erected to conceal from the visitors the destitution of the regime's disenfranchised, a whole little Yiddish world. A woman at her doorstep gestured to him:

Bist a yid?[9]

He was warmly welcomed into a wretched dwelling. They spoke in Yiddish, François muddled through. The father, member of the Party, laid out for François the philosophy that helped them carry on:

Moses never made it to the Promised Land, but he knew his people would get there someday; here, it's the same, we won't get there, but our children will.

The mother, for her part, is more critical. She whispers in François's ear:

The Russians are all thugs, every one of them!

Moscow was all well and good, but François hoped to be able to see the country as he wished — which, at first sight, seemed all but impossible, for the delegations were compelled to stick to the official agenda, fully programmed, and the "youth of all countries" were told they are not allowed to go beyond thirty kilometers outside Moscow. Fortunately,

---

[9]Are you a Jew?

Pierre Aigrain, in his role of guardian angel, had provided him with a particularly glowing letter of recommendation. François sought out the cultural attaché of the French delegation who passed on the letter to the appropriate persons, thereby earning him an invitation to Leningrad by the USSR Academy of Sciences, which allotted him a royal *per diem*, a chauffeured vehicle, a stay at a hotel with an international phone line — very rare at the time — and just for good measure, as much caviar as he could consume!

He made the trip by train, in first class. A man who turned out to work for the KGB[10] was in charge of accompanying him everywhere, a fellow that François was able to make a deal with, for the agent wanted to go visit his parents, which he would normally be unable to do when on a "mission." He left François at the hotel, with full access to the car and driver. So here was François in Leningrad, free as a bird.

He crisscrossed this superb city, visiting the Hermitage, and his driver took him to see the Peterhof Palace, the "Versailles of Saint Petersburg," located on south shore of the Gulf of Finland.

By the time the KGB man returned, a few days later, François was thinking about heading back to Moscow when he realized that, by now, Festival must be over! How was he going to get back to Belgium? But wait, didn't Leningrad have a harbour?

The devoted KGB agent found him a place on a ship leaving for Le Havre, where Esther would come pick him up in her Citroen 2CV.

And that was how a very instructive adventure came to a happy end.

François, who had always harbored a certain sympathy for Communism, returned from the USSR somewhat disenchanted.

In addition to theoretical doubts he had always felt, he was shocked and dismayed at the overt hypocrisy that prevailed in Russian society, the constant surveillance, the persistent anti-Semitism. And this disenchantment would only grow stronger. The crushing of the Hungarian uprising in 1956 plunged many a "fellow traveler" into serious self-doubt. The invasion of Prague twelve years later, followed by the hardly inspiring destinies of Mao's China, of North Korea, of Cambodia and the Khmers Rouges, Ceausescu's Romania ... it was a lot to take.

---

[10]The Soviet state security agency, literally, the Committee for State Security.

François was growing increasingly skeptical. His mistrust of the world around him was deepening, as was his radical refusal of the prevailing "models."

He was greatly relieved to return to his elegant, enigmatic equations.

But one question was still nagging him: was there really such a difference between ideologies? Did they not all rely, in the end, on an appetite for power? And did they not all inevitably lead to violence?

Today, as I write these lines, I figure that the tide has turned significantly. It is important not to be fooled by our political systems, and it is undoubtedly reasonable to always put things into perspective, but a new urgency is now detectible, one that resembles, if I am not mistaken, something that we thought we would never again witness. As Bertolt Brecht says,

"The womb is still fertile from which the vile beast emerged."

We are seeing the surge of brazen shifts among regimes openly advocating racist practices, anti-Semitism, contempt for human rights, ethnic supremacy, and class entitlement.

Is relativism still appropriate today, in 2023?

As our conversation progressed, the subject was inevitably raised, and I could sense in François a fire that had never quite been extinguished, the fire of revolt, resistance, and solidarity in the face of barbarity.

II.3.2. *Double agent!*

François's carefree wanderings during his trip to the USSR in July 1957 would not be inconsequential. Upon his return, there was an attempt to recruit him as a double agent!

It so happened that the KGB agent had a friend who worked in the Soviet embassy in Brussels. François was summoned to appear before Belgian Security, and a few weeks later, Pierre Aigrain was questioned about François by Interpol! Aigrain fought back unhesitatingly: *Come now, François is anything but a spy, he's a brilliant young physicist, above all suspicion*! No one would doubt the word of a man of Pierre Aigrain's stature and reputation.

At this point, François was starting his compulsory military service as a private, having refused to be an officer. He was working at the Royal Military School for the Air Force. There, he was contacted by someone in Security who, reassured most probably by what Pierre Aigrain had said, and hoping to recruit François by promising him a car as a gift, started out by simply asking him what actually happened in Moscow and Leningrad. François very casually explained that he had simply wanted to tour the beautiful city of Leningrad at his ease. They chatted a bit, a certain affinity emerged between them, and François confided that this was the absolute truth, that he was incapable of lying. And then, as if prompted by François's candor, the Security guy confessed that he had always dreamed of being a writer. And this was where François's improbable career as a secret agent came to an end.

But later, when François was invited to pursue his research in the United States, the Soviet incident would resurface. Getting a visa for the US was far from a done deal!

L'école buissonnière: A Sputnik in Belgium

Autumn 1957 was a high point for the USSR: on 4 October, a small, beeping metallic sphere was quietly launched into orbit around the Earth. The whole world was amazed, with the exception of the thoroughly disconcerted United States.

Among François's merry band of friends, the reaction was instantaneous: the perfect occasion for a prank. What if we claimed that the Sputnik landed in our little Belgium? François was all in from the start, given his natural inclination (recall his Uncle Leon, grand prankster before the Almighty) and for the sheer delight in participating in something even a little subversive.

Among the other participants were the highly imaginative Ben Tursch, Georges Miedzianagora, François, and later on Luc Somerhausen.[11]

---

[11] In law school at the time, Luc Somerhausen was one of the architects of the Front du Nord, a Belgian network linked to the French Jeanson network that supported the FLN (Front de Libération Nationale) during the Algerian War (1954–1962). Under the nom de guerre "Alex," he helped militant Algerians who were being persecuted in France to secretly cross the border into Belgium, where they were housed, provided with forged documents, all with the help of the "suitcase carriers," those militants willing to provide logistical help to Algerians. He also worked with a collective of Belgian attorneys who would defend FLN militants in French courts. Faithful to his idea of justice, he would become a judge, and later, the vice president of the Labor Tribunal of Brussels (Colette Braekman, *Le Soir*, 12 March 2012).

With the complicity of an assistant to Professor Kipfer, they "borrowed" the scale model of Professor Piccard's famous bathyscaphe[12] suspended from the ceiling of a gloomy corridor of the Faculté des Sciences: this metallic sphere could easily pass for the Sputnik. Inside it, the pranksters hid a cryptic message addressed to the Belgian gendarmerie: "My ass is a dragonfly." This was in fact the first line of a much-loved student anthem:

Mon cul est une libellule
Qui s'en va chaque matin
Voltiger dans la lagune
Pour y faire des pets marins[13]

They also left inside the cabin a text in Russian that made absolutely no sense, as well as dry ice, which caused the outside of the bathyscaphe to form a frozen layer.

The thing was then left on the sly in the early hours in a trash dump in Nivelles, where one of the pranksters discretely poured gasoline over it, setting off a spectacular cloud of fire around the layer of ice, attempting to recreate the idea of an "object" that had passed through the stratosphere. At the same time, the accomplices stirred up the neighborhood, halting traffic and calling the police.

A "witness," Georges Miedzianagora as it happens, announced the unbelievable discovery to the Belga Agency which phoned the Observatory director. Reactions ran riot, and the press and gendarmerie were quickly on the scene. Local residents were getting out of control, journalists and security officials had no trouble gathering a wealth of witness reports, each one contradicting the other. One saw a fiery arc shoot across the sky, another saw a bright green glow …

The gendarmes were reluctant to open the device, and called in the army. Speculation was rampant: an accident or a NATO sabotage? All the radio stations chimed in with interviews of the characters who just

---

[12] Auguste Piccard (1884–1962), Swiss physicist, hot-air balloonist, deep-sea diver, professor at the ULB. His bathyscaphe made it possible to explore the deep marine trenches. He inspired the Hergé character of Professor Calculus in the Tintin series.

[13] My ass is a dragonfly/That flies away every morning/Hovering over the lagoon/To let fly seaworthy farts.

happened to be strolling around Nivelles that day. Georges Miedzianagora launched into a very scientific-sounding speech, warning the population of the air column above the place where the object fell to earth.

The city of Nivelles organized a conference where a very serious meteorologist came to speak of climate change; making introductory remarks were two guests, the physicist François Englert and the philosopher Georges Miedzianagora, who somehow managed to keep a straight face, even when Georges sets out on a lengthy tirade in comic opera Russian, which François then "translated."

The following day, the truth came out: the whole thing was a prank and pranksters were singled out. Homo belgicus, always a good audience, had a good laugh, and not without a certain sense of pride. The Chancellor of the ULB, Henri Janne, invited the merry band over to celebrate, uncorking a bottle of champagne. He was delighted. The Russians, not so much …

II.4. A Doctorate While in Uniform

In June 1958, François received his degree in Physics. And was daddy to his second daughter, Anne.

As for his doctoral dissertation, he would write it while doing his military service.

By then, François was 25: married with two children, uniformed serviceman, impassioned by the latest scientific conquests, and eager to delve into resolving the enigmas he encountered on a daily basis.

His time in the army provided the chance for him to relish unreservedly the idiosyncrasies and profoundly paradoxical nature of that institution where discipline was supposed to reign supreme, but where orders, as we shall see, can be easily circumvented, and at times, rather surrealistically.

You would find the brass trying to impress the rank and file, throwing tantrums and threatening, but usually ending up looking the other way, out of resignation or good sense; MPs assigned to guard duty playing cards, silently so as not to wake up the young recruits they allowed to sleep on their premises; antiquated technical services but a well-stocked library. François discovered an institution where, if you were clever, you could spend a relatively painless time, if not an exciting one.

In the military, academics were automatically ranked as officers, but François opted to be a simple foot soldier, because his tour of duty would last only 12 months instead of 15. He was appointed to the Air Force, and would spend a few weeks of basic training in Saffraanberg, Limburg, at the École Technique de la Force Aérienne.

There, anyone with a graduate degree who had opted to be a lowly private was considered with certain respect by the higher ranks, who were impressed by their superb disdain of the army's highly hierarchical system. The instructor, on the other hand, set the tone right away: "Here with me, we don't joke around!" To the battalion's few "highbrows," he liked to show that he had a sense of humor: "You eggheads, you understand at least when I say: 'To the rear, march!' It's like forward march, except in the other direction!" He also liked to recite heroic verses of his own: "Once you've had enough of me, you won't be such a sight to see."

But despite all his bluster, he wasn't as strict as all that. Every morning, they were all required to attend flag-raising, where François was never present. He'd rather go out for a walk. Summoned before the higher-ups and threatened with sanctions, he managed to get out of it by claiming he had migraines.

He took part only once in target practice. A large transport vehicle took the battalion out to the firing range some 60 km. from the school. Upon arrival, they realized they'd forgotten the ammunition. When the exercise finally got underway, François, thanks to all the time he had spent at the Brussels fairgrounds, proved an excellent marksman. Another conscript warned him that he'd better stop showing off, since they were liable to induct him into the elite sharp-shooter corps! François heeded the advice and discharged as far from the target as he could, which earned him a look of contempt from the overseeing officer.

After his training at Saffraanberg, François took up residence at the Royal Military Academy in Brussels. He was assigned to Telecommunications, where he had occasion to assist an officer who was preparing a dissertation on Quantum Field Theory.[14]

---

[14] A field is an object which, unlike particles which are localized, is defined in a region of space (e.g., a magnetic field). We will be addressing Quantum Field Theory in more detail in Episode V.

François had done his final Polytechnique work on antennae, so he was ordered to check the measurements of a waveguide.[15] He soon realized that the military personnel don't really understand what they are doing in this laboratory and decided to interrupt the experiments. His superior, vexed at first, finally dropped the matter. Next, they ordered him to cobble together a doorbell. And all in keeping with his skill set!

But he did find some good in the institution, after all. Two things would allow François to move forward in getting his dissertation written:

- The Royal Military School had a remarkable scientific library that subscribed to all the important journals. The civilian commander, Charles, an excellent physicist by trade, freed up a large table for Private François in an empty room that would become his office. Far from the noise and temptations of civilian life, François completed his dissertation in record time!
- Moreover, he had the great good fortune of having as superior the officer whose dissertation he helped to write. By way of thanks, the officer granted him open-ended permission to leave the barracks. François was free to come and go as he pleases! This allowed him to spend afternoons with Jules Géhéniau, professor of Physics and Mathematical Physics at the ULB, who agreed to direct his dissertation.

The professor lived with his charming companion, a cat named Sputnik, and a remarkable wine cellar. He had a deep understanding of quantum physics and general relativity. Like Professor Van den Dungen, Jules Géhéniau was active in the Resistance during the war and organized underground classes when the ULB was shut down.

François's dissertation had a lovely title: *Behavior of a small quantum system immersed in a weakly dissipative medium.*[16]

---

[15] A waveguide is a device designed to confine and direct the propagation of electromagnetic waves.

[16] A dissipative medium is one in which a system evolving within it can dissipate energy in the form of heat.

It was François who proposed the subject, inspired by his articles published around the time he was Pierre Aigrain's assistant, in particular, his quantum study of electron behavior at thermal equilibrium. Using Quantum Field Theory (Section V.2), François showed that in a black body,[17] the interactions between the electrons of the material and the photons[18] present inside modified the mass of the electron.

Once again, François deployed a clever mathematical manipulation to elegantly formulate a generalization of a theory (plasma oscillation[19]) that connected with the concerns of Robert Brout, an American, whom he had not yet met at this time, but who would become one of his most faithful brothers in arms. Robert Brout was the author of a highly refined article on the subject that greatly impressed François. Later, in the United States, Robert Brout and François would pursue this study, among other common interests.

Professor Géhéniau, who very kindly assumed the role of dissertation director, and François, who, as always, liked to work as he pleased, came to a genial understanding. The doctoral candidate and his director would spend only an hour discussing the "small quantum system." Much more time was spent enjoying fine wines and marveling at the cat Sputnik's strange talents as she displayed an uncanny aptitude for returning balls tossed at her, while brilliantly executing the role of goalkeeper, posted between two chairs. Not only could Sputnik catch and return the ball but she always returned to her goalkeeper position.

Jules Géhéniau, like Paul Glansdorff, occupied a special place in the life of the brilliant doctoral student. They would remain friends until Géhéniau's death in 1991.

François defended his thesis in September 1959, and passed, Summa Cum Laude, to become a Ph.D. in the Sciences.

---

[17] A black body is an object that absorbs all the electromagnetic rays it receives. With light totally absorbed, the object is called "black."

[18] We shall be dealing with photons very soon, and in depth (Section III.3).

[19] Rapid oscillations of the electron density in conducting media.

His dissertation was the subject of a publication in *Nuovo Cimento*[20] and would also provide the basis for two other articles[21] in the *Journal of Physics and Chemistry of Solids* and in the *Bulletin de la classe des Sciences de l'Académie royale de Belgique*.

At this point in his career, François was a seasoned researcher with several published articles under his belt, and whose work had produced solid results.

He was yet only twenty-seven years old.

He had more than made up for the time "lost" at the Polytechnique.

In the years to follow, François would meet a number of physicists more or less accomplished than himself, often his elders.

This impression of being "the youngest" would persist in his mind for quite some time, and as the years passed, he would always be astonished to observe that the "others" were not automatically "older" than him; a growing number, in fact, would be far younger.

His status as the youngest of the family, which of course never changed, would continue to plague him …

L'école buissonnière: A Village in Andalusia

> *Les plus désespérés sont les chants les plus beaux,*
> *Et j'en sais d'immortels qui sont de purs sanglots.*[22]

> Alfred de Musset (*L'Allégorie du Pélican*)

In 1958, François and his friend Georges Miedzianagora decided to take a trip. Their destination: South, with no particular place in mind. Esther and her mighty 2CV were along for the adventure, with Georges driving his own car.

---

[20] Englert, Fr. "Interaction entre un petit et un grand système Hamiltonien effectif: polarisation et absorption," in *Nuovo Cimento*, vol. 10, no. 3, 1958, pp. 560–562.

[21] Englert, Fr. "Comportement d'un petit système quantique dans un milieu faiblement dissipatif," in *Journal of Physics and Chemistry of Solids*, vol. 11, no. 1–2, 1959, pp. 78–91; Englert, Fr. "Renormalisation de masse d'un électron dans un corps noir," in *Bulletin de la Classe des Sciences de l'Académie royale de Belgique*, vol. 45, 1959, p. 782.

[22] The most hopeless of songs are most lovely of all/Immortal, pure sobbing, do I now recall.

The voyagers made a stop in Paris to see Marek Halter, who was currently indulging in oil painting.

They went out on the town every night, talking art into the wee hours.

Clara Halter was suddenly gripped by a desire to buy shoes, and only Madrid seemed worthy as a shopping venue. Off they went to Madrid, then, by way of Blois!

In Madrid, they stopped by the Prado, where Georges, pretending to be disabled, visited the galleries in the comfort of a wheelchair!

Marek and Clara went off on their own; Esther, who couldn't stand Georges, soon abandoned the two buddies and headed back to Brussels. Still, she would write to François, *poste restante*, during the entire duration of their trip through Spain.

Georges and François continued their southerly route, and came upon the lovely Andalusian village of Mijas, a mountain town that, back then, was not yet a tourist destination (sadly denatured today by that industry). They next decided to push on to Morocco by boat, from the port of Algeciras. Unfortunately for them, they arrived at Algeciras during Holy Week: the people's fervor felt both hysterical and sinister, in what was Franco's Spain at this time, and the processions carrying statues of the Virgin were replete with soldiers — testimony to the Church–Army alliance.

The next day, when they were about to board the ferry for Morocco, two police officers intercepted Georges and François. The two friends were sure they were being arrested for blasphemy, because the previous evening, Georges couldn't help making some lewd remark as the Blessed Virgin passed by, but that wasn't it at all. The policemen simply wished to let them know there was someone waiting for them. Surprise! There was Esther and her 2CV! Georges was annoyed, but they made the trip as a trio, a somewhat improvised and chaotic journey, it turned out. In Tetouan, they checked into a shabby hotel where they found a clogged bathtub full of foul-smelling water. The friends decided that it was time to stock up on marijuana, and they entrusted a local kid with five Belgian francs … which bought them two large packages of hashish and a few square wafers of a particularly fearsome mixture! They would bring this bonanza back with them into Belgium without a single problem, using Esther's undergarments as their cache!

It was also in Tetouan where François set foot for the first time in his life inside a synagogue, thanks to a street urchin who served as their self-styled tourist guide. He took them first to visit the old Arab quarter, the Medina, a tangle of interlocking white hovels, and then to the old Jewish quarter, the Mellah, which, according to the guide, was where "the rich Jews" lived, though the visitors noted that their houses were just as tiny and miserable as the ones in the Arab quarter. The only difference: the Jews didn't have goats gamboling through the streets.

François's keenest memory from Morocco was an emblematic image: a fellow riding a donkey, relaxed, his legs dangling, and alongside him, his wife carrying a heavy load …

Return to Andalusia.

The village of Mijas, near Malaga, appealed greatly to François and Georges. It would become a rallying point for all the friends. At that time, houses and property could be bought for a song. François rented for a laughably low amount a mill with a swimming pool, and a "little maid." Paul Glansdorff rented a grand house! The friends all descended on Mijas: Joseph Katz and his new partner, Georges, Marek and Clara, the couple "Bob and Bobette," and the philosopher Pierre Verstraeten, with whom Georges would inevitably come to blows. Later, Robert Brout would vacation there with his wife Martine.

Such a concentration of rebellious individuals gifted with a fertile imagination would not remain peaceful for long. The atmosphere in Mijas gradually deteriorated. Trivial incidents exploded into wild scenes. Georges had a knack for starting fights. Innocent evenings at a local eatery ended in violent shouting matches!

During these trips, Georges revealed other unexpected facets of his personality. During a car trip to Ronda to attend a bullfight, for instance: the road was narrow and winding, along a ravine. François came out suddenly with, "So, I guess all you have to do is make a sharp turn, and it's over!" Georges, who didn't think this was funny at all was gripped by a genuine panic attack.

Seemingly dormant volcanoes that suddenly explode in a burst of uncontainable emotion, such is the image that escapees from Nazi terror reflect. Their childhood wounds never heal and persist into adulthood. How could they not?

François during one of our interviews,

My brother Marc and I are very different. Marc is a man who has certainties, where I have none. I am a mixture of joie de vivre and despair, I have moments of enthusiasm, but the shadow of failure and catastrophe loom at all time. When anyone asks me if there is an episode in my life that I would enjoy reliving, I come up empty-handed. No, I would not like to relive my life.

Episode III

A Stroll through the Land of Physics

There was a time when all cultured persons owed it to themselves to know as much about art as about the sciences of his era. For a century now, that symmetry has been broken, alas. People can consider themselves "cultured" today without knowing anything about the expansion of our universe or about the mystery that two theories as powerful as quantum mechanics and general relativity are not everywhere compatible.

Guy Duplat

This observation by Guy Duplat[1] is highly relevant, and this episode will allow the uninitiated reader to get familiar with what science tells us about our universe today.

III.1. Physics ... A Late Blooming

This terror, then, this darkness of the mind,
Not sunrise with its flaring spokes of light,
Nor glittering arrows of morning can disperse,
But only Nature's aspect and her law,
Which, teaching us, hath this exordium:
Nothing from nothing ever yet was born.

[1] Guy Duplat is an engineer-physicist and cultural collaborator at the daily paper *La Libre Belgique*, currently sold under the title *La Libre*.

> *Fear holds dominion over mortality*
> *Only because, seeing in land and sky*
> *So much the cause whereof no wise they know,*
> *Men think Divinities are working there.*

Lucretius, *De rerum natura* (I, 146)

There isn't much to add to this extract from the long poem *On the Nature of Things*, written in the first century BCE by Lucretius, disciple of Epicurus.[2] We *homo sapiens*, over all eras and civilizations, have never ceased our search to uncover the mysteries surrounding us, threatening us, thus revealing our own inherent weakness. We once called upon supernatural powers to come to our rescue, but their responses fell short, and a vision of the world that could dispense with the gods would finally prevail.

Physics seeks the intelligibility of nature in rational terms. Or more precisely, today's fundamental research in physics attempts to interpret all phenomena, in their diversity, as particular manifestations of general, universal laws, a handful (or even down to a single law that would encompass all of them) that would be experimentally verifiable. Thus, for example, the trajectory of a massive object, the revolution of planets and satellites, the movement of the tides can all be explained and even described quantitatively in the language of mathematics by the general law of gravity formulated by Newton in 1687.

But the notion of a world regulated by a small number of general laws, though prepared over centuries or even millennia of speculation and observation, is still surprisingly recent in human history. It emerged in Europe during the Renaissance. It proved a veritable scientific revolution that would give rise to unprecedented knowledge and technology. We shall address this extraordinary intellectual adventure.

---

[2]Lucretius (98?–55 BCE). *De rerum natura* (On the Nature of Things) contains the most thorough exposé of the philosopher Epicurus (341–270 BCE), a prolific thinker but only three of whose letters have been found. As for the text of "De rerum natura," it has come to us via 9th century manuscripts and the burnt remains of a papyrus scroll found at Herculaneum. The translation is that of William Ellery Leonard, E.P. Dutton, 1916. Lucretius, De Rerum Natura, Book I, line 146 (tufts.edu).

III.1.1. *The ancient world*

Since time immemorial, and in the four corners of the globe, learned people have observed and carefully measured the movements of celestial bodies observable with the naked eye and have established highly precise calendars and almanacs.

The most ancient traces of what we would call science today date back several millennia and concern civilizations that occupied the fertile plains of China, India, Egypt, and Mesopotamia.[3] Star-gazing played an important part, as did mathematics, which was remarkably well developed in Babylonia in the second millennium BCE.

This knowledge gradually spread throughout the Mediterranean world. It is important that we not forget what we owe to these distant precursors, but today in Europe, we see ourselves above all as children of ancient Greece, for our schooling is largely based on the Greek model. And if we have chosen to outline Europe's scientific venture by starting with the Greek world, it is because, as we shall see, what we call modern Physics was born by destabilizing those foundational principles.

The astronomers of antiquity of course wondered about all the whys and wherefores of what they were observing. Thus, numerous models of the universe emerged, all strongly tinged with the mythological beliefs and philosophical or religious dogmas of the age.

A Greek citizen's vision of the world at the time of Homer[4] was probably that of a flat earth, surrounded by water and topped with a closed dome where luminous specks and celestial bodies moved as if by magic. (A recent statistic revealed that many young Americans share this vision, one that seems to be gaining credence worldwide.)[5]

It was not until the 6th century BCE that the spherical model of the Earth prevailed, a hypothesis already posited by Babylonian astronomers almost a millennium earlier.

---

[3] We should also mention traces of the same nature concerning the pre-Columbian civilizations of Latin America.

[4] Eighth century BCE, most probably.

[5] See, for example, the book by Lee McIntyre of Boston University, *Defending Science from Denial, Fraud and Pseudo-Science*, MIT Press, 2019.

For Pythagoras (born around 580 BCE), and later Plato (born around 428 BCE), the Earth was by definition *round, immobile at the center of the world, and very large.*

Aristotle (born around 384 BCE) would provide the first proof that the Earth is round, notably by observing eclipses.

This geocentric model of the Cosmos, discarded since, proved nevertheless extremely fruitful.

In ancient Greece, there developed a powerful mathematical astronomy based on Euclidian[6] geometry (dating back over 2,300 years, it is still today the basis for teaching geometry). Applying this geometry, Eratosthenes, in 230 BCE, succeeded in determining with astounding precision the circumference of the Earth!

According to Ptolemy's[7] famous treatise, the Almagest, the earthly globe occupied the center of the world, around which the sun completed diurnal and annual revolutions, the planets gravitated, and the stars were fixed points on a sphere in rotation. This model made it possible to predict with relative accuracy the movements of the stars, but led to a highly complicated description of celestial movements, and especially prevented imagining that the phenomena observed in the heavens might be governed by laws equally applicable to phenomena observed on earth. Moreover, finding a dynamic explanation for celestial movement was not the astronomers' main concern; in the absence of the now familiar concept of interaction among distant material bodies, how could they have glimpsed even a hint of an explanation for the extraordinary ballet of the stars that was taking place over their heads, in a world where astrology and astronomy meant the same thing, where almanacs could predict both solar eclipses and the fate of kings, or the moods of the gods, where cosmology[8] and superstition mutually reinforced one another?

It would take the Galilean revolution to understand that laws of physics exist and that these laws had a universal status.

---

[6] Euclid (probably around 300 BCE).

[7] Ptolemy (100?–168?), Greco-Egyptian astronomer and astrologer, Roman citizen.

[8] The science of laws that govern the universe. Cosmology covers a much broader field than astronomy, which is devoted to the observation of the sidereal. The point of cosmology is to write the history of the universe.

III.1.2. *Renaissance and Galilean revolution — The principle of inertia*

Copernicus (1473–1543), followed by **Galileo** (1564–1642), would turn the cosmology inherited from the Greeks on its head.

Nicolas Copernicus put the sun at the center of the universe, and around this fixed point revolved the Earth, just one planet among other planets. In 1512, Copernicus posited his heliocentric model[9] in a short treatise that had to be circulated covertly among his friends. For thirty-six years, Copernicus would refuse to divulge it, as much out of a self-imposed insistence on scientific rigor as fear of how the Church might react to a notion of the world that deconstructed the Biblical word. It was only in 1543, upon the death of its author, that the work would be published.

From **Galileo**, we would learn, thanks to his improvements of the astronomical telescope (invented around 1608), that the moon's surface is far from smooth, that the sun has spots, Jupiter has moons, and that a cluster of stars forms what we would later call the Milky Way, or the galaxy, ours.

But his most decisive contribution was the principle of inertia that would reveal the universal character of the laws of physics. Galileo formulated the inertia principle somewhat imprecisely, but that takes nothing away from his accomplishment. All his physics was based on the very essence of this principle: if a body is at rest or in a state of motion at a constant speed (uniform motion), it will remain at rest or keep moving at a constant speed unless it is acted upon by a force.

Here, Galileo was not thoroughly defining what he called "uniform motion." It was Newton who would further make clear that what is meant is the "uniform rectilinear motion," called "inertial motion."

In fact, this principle, today verified by countless experiments, posits the impossibility of detecting, through a physical experiment within a system, the movement of the system if this movement takes place in a straight line at a constant speed. The fact that we do not perceive the

---

[9] Several ancient Greek astronomers had already posited the hypothesis of a heliocentric model, as well as certain Indian and Arab astronomers, and later, Leonardo da Vinci.

inertial movement of the train or plane in which we are traveling is not only proof of the inertia principle but expresses a profound idea: it heralds the universality of the Galilean approach for understanding all of nature. This is because this example (a rough equivalent of which Galileo was able to observe on a boat in motion) implies that we ourselves constitute a physical system and that the laws of physics apply to all systems, whether inanimate or living.

Accordingly, it was starting to seem clear that this principle, and more generally all the laws of physics, should be liable to be applied both in the heavens and on the earth. And it was commonly agreed that the heliocentric model of the cosmos posited by Copernicus was more natural, with its simpler description of the movement of celestial bodies that opened the way to the Newtonian explanation of their trajectories around the sun. A reworked Copernican model[10] would soon be adopted by the state-of-the-art astronomers. Unfortunately, since it contradicted the literal reading of the Bible, heliocentrism would attract equally fierce and aberrant reactions from the Catholic Church at the time.

By the universal character of the inertia principle he had discovered, by his many observations of the heavens, and by his many experiments involving objects in motion on the earth which he described in mathematical terms, Galileo ushered in the era of scientific astronomy, and more generally, the era of rational physics. This field would develop with astounding speed. His intuition as to the universality of the laws of physics is no longer a matter of debate. Since the early 20th century, we have detected from locations on the Earth the electromagnetic radiation emitted or absorbed by billions of celestial objects, and the resulting analyses are conclusive: in the whole universe, all the atoms observed are identical to those that comprise us, ourselves, and our planet. The heavens and the Earth are offspring of the same brood.

---

[10] For Copernicus and Galileo, the trajectories of celestial bodies around the sun were circular, though we now know they are elliptical, as demonstrated by Johannes Kepler (1571–1630), and as imagined already in the 11th century by the Andalusian astronomer El Zarqali.

Many outstanding men and women have contributed to developing what we call today's "modern physics." At the decisive stages of this evolution, the names forever associated are

III.1.3. *Newton, Maxwell, and Einstein*

Isaac Newton (1643–1727) formulated his general laws of gravitation that govern the movement of all massive bodies[11] both in the heavens and on earth. He envisioned a universe made up of tiny bodies interacting through forces that trigger accelerations and that determine their trajectories. In a way, these bodies foreshadow the elementary particles that are the ultimate constituents (at present) of all the known objects in the universe — their structure will be discussed in Episode V.

Newton's first law revisits Galileo's notion of inertia:

"The innate force of matter is a power of resisting by which every body [...] endeavors to persevere in its present state, whether it be of rest or of moving uniformly forward in a straight line. [...] Inertial motion having no physical effect and demanding no cause, any modification of the path of a body, whatever its nature, can only be caused by the action of an impressed force."

This altogether novel notion of a force coming from an external source (the sun, for instance) and which can act from a great distance on the other bodies (the planets) defines **a long-range interaction**. We could ask ourselves the question, then: might there exist in the universe other long-range interactions than those related to gravitation?

The answer is yes: electromagnetic interactions, inextricably linked to the name of their "discoverer," James Clerk Maxwell.

James Clerk Maxwell (1831–1879) formulated the general laws of electromagnetism, i.e., the laws governing the properties of electrically

---

[11] We have all come into contact at school with the universal law of gravitation that allows us to calculate the attractive force exerted between two masses m and M separated by the distance R. This force is proportional to the product of the masses and inversely proportional to the square of the distance separating them, with the proportionality factor G, the gravitational constant. Thus, F = G. (m.M/R^2).

charged bodies, their coupling[12] to electromagnetic waves and their properties. The laws are expressed in terms of a new notion, that of electromagnetic field,[13] an object defined in a region of space where a new type of long-range interaction prevails, electromagnetic interaction. While for Newton, light was a beam of particles rapidly moving in straight lines, Maxwell showed that light is a particular kind of electromagnetic wave associated with an electromagnetic field.

Maxwell's electromagnetic theory covers all electric, magnetic, and light phenomena.

Albert Einstein (1879–1955) made several fundamental contributions to the development of physics:

He extended the validity of the inertia principle to electromagnetism, which was not trivial, since Maxwell explicitly introduced the speed of light in his theory. And because the inertia principle then requires that the speed of light not be affected by a system's inertial movement, notions of time and space had to be modified. Einstein's answer was a theory where the speed of light is constant, its value being, in fact, the maximal velocity of a signal in the vacuum.[14] This is the theory of special relativity (1905), confirmed by countless experiments in its field of application. It also implies that the elementary particles that make up the universe (Episode V) are as follows:

- either massive particles that can never reach the speed of light,
- or massless particles whose speed in a vacuum is always equal to the speed of light.

It was in his 1905 publication that the famous "Einstein relation" $E = mc^2$ (E = mc^2) appeared, establishing the equivalency between mass and energy. It means that a massive particle has intrinsic energy even

---

[12] Coupling is a measure of the interaction.

[13] A charged body creates in the space surrounding it an electric field. When it moves, it creates in the space surrounding it two coupled fields, an electric field and a magnetic field: this is an electromagnetic field, which propagates as electromagnetic waves.

[14] The speed of light $\approx 3.10^{10}$ cm/s, or 300,000,000 meters per second. The equal sign $\approx$ means an order of magnitude and not an exact value.

when at rest. In this relation, **m** designates the "mass at rest," **E** is energy, and **c** is the speed of light. This relation remains valid when **m** designates the "mass in motion", while in Newtonian physics, the mass of a material particle is a stable feature, in the theory of special relativity, the mass increases with the speed of the particle, but this relativistic effect only matters for speeds approaching the speed of light. At that speed, an object's mass becomes infinite, which explains why a massive particle can never reach that speed. As for zero-mass particles, this phenomenon does not apply, for they are never "at rest" and are always in motion at the speed of light c.

He generalized the Newtonian theory of gravity by formulating the theory called general relativity[15] (1915). This theory predicts the effect of gravity on light; it tells us that the attraction we observe between two bodies is due to a distortion of space and time caused by the presence of these bodies. The force of gravity is hence interpreted as a "bending" of space-time, a bending that is itself produced by the presence of mass and energy. This theory differs considerably from Newton's for intensive gravitational fields where heretofore unknown objects appear, those we know today as black holes, gravitational waves, and the like; but it does agree with Newton's theory of gravity in the area of slower speeds as compared to the speed of light, i.e., in the more practical area of our daily life. Extensively verified through experience, it has become an essential element in modern astrophysics and has opened up to scientific investigation the study of cosmology which could not be accurately approached within the framework of Newtonian physics.

Generally speaking, the tag "relativist" when applied to an object means that this object is moving at a speed approaching or equal to the speed of light.

In 1905, Einstein showed that the photoelectric effect,[16] observed for the first time by Antoine Becquerel in 1839, also described in 1887 by

---

[15] General relativity satisfies the inertia principle, but only locally. In fact, around a point and at a given moment, we can always find a system of reference that cancels out gravitation and where the inertia principle remains valid.

[16] The emitting of electrons by matter (a metal, in general) exposed to electromagnetic radiation.

Heinrich Hertz, but remaining mysterious, could not be understood so long as light is considered as being simply a wave. Maxwell's electromagnetic waves are in fact made up of a large number of zero-mass particles whose speed is always equal to the speed of light. In 1916, these "light quanta"[17] were named "photons"[18] in certain publications, a designation that was unanimously adopted around 1926. They played a crucial role in the field of quantum physics that developed during the first decades of the 20th century.

This is what Newton's rapidly moving particles had become, conveyed in large number within Maxwell's electromagnetic waves under the name of "photons."

III.2. The Atom and the Emergence of Quantum Physics

Nothing exists except atoms and empty space; everything else is opinion.

Democritus

Moreover, if there be not a least thing, all the tiniest bodies will be composed of infinite parts, since indeed the half of a half will always have a half, nor will anything set a limit. What difference then will there be between the sum of things and the least of things? There will be no difference; for however completely the whole sum be infinite, yet things that are tiniest will be composed of infinite parts just the same. And since true reasoning cries out against this, and denies that the mind can believe it, you must be vanquished and confess that there are those things which consist of no parts at all and are of the least nature.

Lucretius — *De Rerum Natura,* I 615

---

[17] Quantum, Latin for "defined quantity," plural "quanta," represents in physics the smallest part, whether for energy, charge, etc.

[18] From the Greek *phos*, meaning "light."

Old-timers like Democritus and later Lucretius already had the premonition that matter was made up of a certain number of basic elements, but these visionary convictions did not rest on what we could call a scientific approach, nor did they give a clue as to how macroscopic bodies relate to their constituents. Democritus was a philosopher and Lucretius a poet-philosopher, both curious and subtle observers, but experimentation was not really accessible in their world. It was Lavoisier (1743–1794) who showed by experimenting that matter could be broken down into molecules, and molecules into atoms. At that time, it was thought that the ultimate element had been discovered: the atom (*a-tom* means non-divisible).

During the second half of the 19th century, the evolution of science would take place under the spell of the atom.

But one should know that for a long time, many physicists continued to doubt its very existence. It was the theoretical explanation of Brownian motion[19] given in 1905 by Einstein that put such doubts to rest.

In 1869, Dimitri Mendeleev, a Russian chemist, came up with the idea to classify all known atoms in a table according to their mass and chemical properties. His famous Mendeleev Table revealed a periodicity impossible to interpret if one kept to the notion of an indivisible atom.

The discoveries of X-rays in 1895, of radioactivity[20] in 1896, and of the electron in 1897 ask the question outright: aren't atoms themselves composite objects as well? How else can one explain spontaneous emissions of particles by matter as happens in radioactivity?

Various models of the atomic structure were suggested.

In 1911, the physicist Ernest Rutherford from New Zealand noticed that when a sheet of mica was bombarded with some positively charged

---

[19] In 1827, the Scottish botanist Robert Brown noticed that small particles in suspension in a liquid are subject to continuous stochastic movements. The suggested explanation was that liquids are made up of molecules in constant motion due to thermal agitation and against which the suspended particles are bumping. Einstein derived the theory which would establish irrevocably the existence of atoms.

[20] Radioactivity is the physical phenomenon whereby atomic nuclei spontaneously transform into other atoms (through disintegration) while emitting particles. It was discovered in 1896 by Henri Becquerel for uranium, and very soon confirmed by Pierre and Marie Curie for radium.

particles, these were strongly deviated. He then imagined the atom made up of a positively charged nucleus containing most of the atomic mass, and, separated by a void, negatively charged electrons orbiting around it, like planets around their star.

This "classical" model no longer fits with what we know today about the atomic structure, but it had its usefulness, as is often the case with half-baked concepts in the course of the exciting pursuit of the real nature of things that physics is.

One suspects that emissions of particles or light indicate exchanges of matter and energy between the atom and its environment and most probably imply interactions between orbital electrons and external elements (e.g., a light beam). But what exactly takes place?

Actually, Rutherford's classical model does not work because, due to the electromagnetic radiation emitted by the electrons while orbiting the nucleus,[21] the atom would be totally unstable. It is quantum physics that will ultimately provide the explanation of its stability. In order to understand this, we cannot fail to mention the amazing intellectual adventure of the great physicist Max Planck (1858–1947).

In fact, Max Planck was slow to rally to the cause of atomism. However, by 1894, he was very interested in the electromagnetic radiation of black bodies.[22] This radiation (thermal radiation) exhibits a very singular spectrum[23] and its intensity depends solely on the temperature inside the black body.

At the time, the interpretation of these experimental data was extraordinarily difficult. Planck came forward with the daring hypothesis that exchanges between matter and electromagnetic radiation do not happen in

---

[21] According to the laws of classical electromagnetism, an electric charge moving in accelerating motion emits an electromagnetic field; the orbiting electrons thus emit an electromagnetic field, thereby losing energy, and making the atom totally unstable.

[22] A black body absorbs all the electromagnetic radiation it receives, (let's imagine, e.g., an oven with thick opaque walls and a very small aperture.). This absorption induces an increased thermal agitation inside which generates the emission of an electromagnetic radiation called thermal radiation.

[23] In a graph showing radiated intensities as a function of light waves, we can see that to each particular temperature corresponds a particular curve: the Planck curve or Planck distribution for the given temperature.

a continuous way. He held that matter and light have something discontinuous in their nature. Absorption or emission of radiation by a black body could happen solely through a discrete process, by exchanges of "energy grains" that he named "energy quanta," and whose energy $\mathbf{E}$ and frequency $\mathbf{v}$ (*nu*) are coupled by the relation $\mathbf{E} = \mathbf{h}. \mathbf{v}$ where $\mathbf{h}$ represents a constant value which he managed to calculate.

Thanks to the introduction of this constant $\mathbf{h}$ which would one day bear his name, Planck managed to derive his famous "black-body formula," i.e., the distribution law of the energy radiated at a given temperature, a law which is in perfect agreement with the experimental data! An outstanding result, born from the favorable encounter between an awesome "thought experiment"[24] and a deep understanding of thermodynamics.[25]

The idea of the "quantization"[26] of radiation is now in its embryonic stage. But Planck doubted his own hypothesis. The notion of "discontinuity" troubled him. He didn't know how to interpret this property which would go on to revolutionize physics. And despite the numerous experiments that would confirm his hypothesis, he did not try to dig deeper. It was Einstein who would come up with the right interpretation, by recalling the "photoelectric effect": the "energy quanta" of Planck were very real and were none other than photons, exchanged when matter absorbs or emits light.

For his contribution to the discovery of quanta, Planck would receive the 1918 Nobel Prize. This high distinction pays tribute, in a way, to the universal status of his constant h, henceforth intrinsically linked to quantum history.

For his discovery of the law governing the photoelectric effect, Einstein would be awarded the 1921 Nobel Prize.

In 1913, the Danish physicist Niels Bohr, inspired by Planck's vision, refined Rutherford's "planetary" atomic model. Defying the laws of

---

[24] Einstein's much loved *gedankeexperimenten*, though already pursued by Galilee: a way to solve a problem by the sole power of imagination.

[25] Thermodynamics studies the effects of temperature and heat on the properties of bodies.

[26] The verb "to quantize" means here "to make compatible with the theory of quanta."

classical electromagnetism, Bohr suggested that electrons do not emit any electromagnetic field while orbiting around the nucleus, and that each orbit corresponds to a particular electron energy, whereas energy exchanges with the external world correspond to electrons jumping from one orbit to the other. These energy exchanges do not happen in a continuous way, but by absorption or emission of "quantas," i.e., photons. Whereas this first "quantum model" allows for an interpretation of the stability of the atom, we will see that the slightly naive notion of "planetary" orbits will soon have to be abandoned.

But the essential feature of the message is there: it is the corpuscular nature of light in its correct quantum description that explains the stability of atoms. Those in the field sense that this is no isolated phenomenon and that all so-called "classical" laws will have to be corrected when they deal with events at levels under a certain threshold. Maxwell's electromagnetism, which does not allow the understanding of atomic structure, will have to make room for its quantum version, "quantum electrodynamics," which we will come to know better in Episode V.

In 1925, two physicists, while attempting to translate into equations the behavior of electrons in the atom, would, quite independently, lay the foundations for quantum physics: Werner Heisenberg, a young assistant at Göttingen University, and shortly later, Erwin Schrödinger, professor at Breslau University.

Werner Heisenberg, excellent mathematician, chose to apply very sophisticated mathematical methods to calculate the electron "orbits" in the current atomic model. He wrote out a first mathematical formulation of quantum physics which correctly accounted for a certain number of observed phenomena, including emissions and absorptions of light (photons) by atoms corresponding to electrons jumping from one orbit to the other.

Erwin Schrödinger, on the other hand, opted to represent an atomic electron by a wave (without having a ready explanation for his choice). He imagined a wave associated to any massive (non-relativistic) particle,[27]

---

[27] The idea was around at the time; it got remarkably developed by the French physicist Louis de Broglie, who received in 1929 a Nobel Prize "for his discovery of the wave nature of electrons."

wrote down the equation describing the quantum evolution of this wave, and determined the possible energy states of the given particle.

Schrödinger's approach soon became very popular with many physicists because of its relative simplicity and the fact that it led to a more intuitive perception than Heisenberg's very abstract approach.

Both approaches were nevertheless equivalent, as established the next year by the British physicist Paul Dirac. He did bring both together in one mathematically coherent theory. His masterly book *Principles of Quantum Mechanics*,[28] where he proposed a general formulation of quantum physics, would become the reference textbook of the new physics.

In 1932, the Nobel Prize was awarded to Werner Heisenberg for "the creation of quantum physics."

In 1933, Erwin Schrödinger would share the Nobel Prize with Paul Dirac "for the discovery of new productive forms of atomic theory."

III.2.1. *Probabilities and uncertainties*

> *Incertitude, ô mes délices,*
> *Vous et moi nous nous en allons*
> *Comme s'en vont les écrevisses,*
> *À reculons, à reculons*[29]
>
> G. Apollinaire, *Le Bestiaire*

It was Max Born who, in 1926, would give the physical interpretation of Schrödinger's equation. For this important contribution, he received the 1954 Nobel Prize.

This interpretation introduces a probabilistic element: Schrödinger's wave (often called wave function) does not represent a material particle, but represents the probability of its presence at a given time in a certain point. The planetary model of the atom was dethroned for good.

---

[28] Dirac, P. *Principles of Quantum Mechanics*, Oxford, Clarendon Press, 1930.

[29] Uncertainty, O my delights/You and I, we take our leave/As do crawdads and prawns/ Backwards, backwards.

A massive particle (a non-relativistic particle) is thus an "object" whose extension in space is to be understood in terms of probabilities. Here is the origin of the famous "uncertainty principle" or rather "indetermination relation" issued by Heisenberg in 1927, which tells us that there is a theoretical limit to the precision with which one can measure simultaneously, for example, the position and the momentum of a particle.[30]

Any improvement in the precision of the measurement of its momentum will imply a lesser precision in the measurement of its position, and vice versa.

This limit has an important consequence: the quantum state of a system represents all available information, and the probabilistic element will have to be taken into account as to the prediction of the results of a measurement.

The intrusion of a probabilistic element in quantum physics, which has been a tremendously fruitful milestone, continues to raise numerous questions even today, which will be discussed in the Epilogue of this book.

This probabilistic element renders the interpretation of quantum physics very problematic. We have the impression that we live in a "classical" world, where objects and their behaviors may be perceived or visualized through analogy. At the atomic scale, it is totally different; you don't see anything that is happening. You'll know about events only through experimental measurements, reasoning, mathematical formulations, and a bold imagination. Quantum physics may often appear paradoxical, because everything there seems so far from our standard conception of the nature of things, but one should know that its mathematical formulation is unambiguous, its experimental verifications are in no way questioned, and that its statements are highly accurate, enabling the prediction of critically important phenomena.

III.2.2. *The 1927 Solvay Conference*[31]

Among the 29 giants of physics participating at the Solvay Conference held in Brussels in 1927 were all the scientists we just mentioned: Max Planck, Albert Einstein, Niels Bohr, Werner Heisenberg, Erwin Schrödinger,

---

[30] For a massive particle, the momentum is the product of mass by velocity: $p = m.v.$

[31] The Solvay Conferences are scientific conferences of Physics or Chemistry held since 1911, bringing together every third year highly eminent scientists. They have branded the history of modern science, leading to major breakthroughs, notably in quantum physics. See bibliography at the end of this book.

Paul Dirac, and Max Born.[32] The theme of the conference was none other than "Electrons and Photons."

This 1927 Conference was no doubt the most famous one, as it somehow endorsed the birth certificate of the "quantum revolution." And it was during this Conference that one of the crucial battles between the various interpretations of quantum physics took place. There is an abundant literature about the passionate discussions where two of the main actors confronted each other: the "Copenhagen school" represented by Niels Bohr and Werner Heisenberg, who had turned the probabilistic element into the keystone of the new physics, versus Albert Einstein and Erwin Schrödinger, who found this element very disturbing.[33]

We will return in the Epilogue to the issues discussed at the time. But let's enjoy, as a foretaste of that atmosphere, this excerpt from a letter from Paul Ehrenfest, close friend to both Einstein and Bohr:

It was splendid for me to be present at the dialogues between Bohr and Einstein. Like a game of chess. Einstein always with new examples. Something like perpetual motion devices of the second kind, but to violate the uncertainty relations. Bohr always searching, out of a dark cloud of philosophical smoke, to find the tools to shatter example after example. Einstein like a jack-in-the-box: springing out again fresh every morning. Oh, that was priceless. But I am almost unreservedly pro-Bohr and contra-Einstein.[34]

III.2.3. *But what is the nucleus made of?*

It is easier to smash an atom than a prejudice.

Albert Einstein

---

[32] Out of the 29 participants, 19 had been or would eventually be awarded a Nobel Prize, including the only woman present, Marie Curie, who would earn two Nobel Prizes!

[33] Diu, B. "Le Congrès Solvay de 1927: Petite chronique d'un grand événement," in *Bibnum.* Online April 1st 2009, htpp://journals.openedition.org/bibnum/846.

[34] Klein, M.J. "Great Connections Come Alive: Bohr, Ehrenfest and Einstein," in J. de Boer, E. Dal and O. Ulfbeck (eds.), *The Lesson of Quantum Theory*, Amsterdam, Elsevier Science Publishers, 1986, p. 338.

Compared with the massless photon, or the "fluffy" electron, the nucleus seems to be a very solid object.[35]

In 1919, Ernest Rutherford discovered the existence in the nucleus of the proton, a particle carrying a positive charge and much more massive than the electron.

In 1932, the British physicist James Chadwick uncovered the existence of the neutron, a particle much like the proton but without any electric charge, hence its name.

Around the same time, Werner Heisenberg suggested an atomic nucleus consisting of protons and neutrons — called nucleons.

With the discoveries of the electron, the proton, and the neutron, and knowing that an atom is electrically neutral, we can imagine it as a small system consisting of Z electrons (charge $= -1$ each), linked by electromagnetic forces to a nucleus consisting of Z protons (charge $= +1$ each) and $N–Z$ neutrons, where N is the total number of nucleons. If one classifies all known chemical elements according to their Z-number, it appears that this classification corresponds to Mendeleev's Table, and we can thus finally hope to understand the observed periodicity.

Throughout this deconstructive journey, an obvious question arises: are nucleons indivisible or not?

In order to answer that question, why not try, for instance, to smash the nucleus by bombarding it with beams of particles?

In the 1920s, it seemed obvious that such an in-depth study of matter will require beams of very high-energy, controllable particles. Various sources are at our disposal: electrical discharges in gases produce ions,[36] heated metals emit electrons, etc. Today, one uses essentially protons and electrons.[37]

These "bullets" will have to be accelerated to increase their individual energies. Several devices have been tested with variable success. In short,

---

[35] The nucleus, which is about 100,000 times smaller than the atom, carries practically all of its mass.

[36] An ion is an atom which has lost or gained one or several electrons; ions are thus electrically charged objects.

[37] In a 1-liter bottle of hydrogen, one has about 50×10^{21} protons and as many electrons, enough to feed a particle accelerator for centuries!

the necessary devices needed to significantly accelerate charged particles are strong electric and magnetic fields, and a very deep vacuum — electric and magnetic fields to accelerate and guide the particles, and a vacuum to prevent the particles from slowing down due to collisions with other particles present in the tube where the beam is circulating.

The first circular particle accelerator was built at UC Berkeley in 1931 by Ernest Lawrence and Milton Livingston. Accelerators will never cease to be upgraded in order to produce higher energies.

Inside an accelerator, the beams of particles travel (in opposite directions) at very high speeds and energies before colliding. One discovers then that the protons and neutrons that make up the nucleus are actually made up of various types of elements that seem unbreakable, at least at the level of energies presently used, thus earning, for the time being, the title "elementary particles."

To give an idea of the superb extravagance of this fundamental research, the LHC at CERN,[38] today the most powerful accelerator in the world, consists of a ring that measures 27 km in circumference, comprising superconductor magnets[39] and acceleration devices, where the particle bullets reach speeds very close to the speed of light (at 99%). Today, in the LHC, they reach about 7 TeV[40] for protons, energies which endow them with a motion-mass 7,000 times their rest-mass (Section III.1). On July 5th 2022, collisions between protons at energies of 13.6 TeV were detected in the LHC, a value never before achieved.

III.3. The Photon, a Little Fellow Well Deserving of a Couplet

First of all, let's recall a few elements of Maxwell's theory of electromagnetism.

In that "classical" theory, light is electromagnetic radiation.

---

[38] The CERN (Conseil Européen pour la Recherche Nucléaire) founded in 1954, near Geneva, Switzerland, aims to discover the ultimate constituents of matter. It is therefore usually named "Europeen Laboratory for Particle Physics." LHC = Large Hadron Collider.

[39] Particularly powerful magnets. We will come back to the phenomenon of superconductivity in Episodes IV and V.

[40] TeV = Tera-electronvolt = 10^{12} eV. Remember Einstein's relation $E = mc^2$.

Everyone has observed at some time, whether at school or elsewhere, the decomposition of visible light by refraction through a prism and the appearance of "all the colors of the rainbow," each color actually representing what we call a "monochromatic wave." Every electromagnetic wave can always be described as a superposition of monochromatic waves. True to its name, each monochromatic electromagnetic wave has two components, one electric and one magnetic, each with its respective field, electric or magnetic. This is a coupled oscillation of an electric field and a magnetic field in a plane perpendicular to the direction of propagation of the light beam.

Since the variations of electric and magnetic fields are linked by Maxwell's equations, let's keep things simple and describe only one of these fields, i.e., one of the components of the wave. And it is usually the electric component that is described when simplifying.

The vibration directions of the electric field can have any orientation in the plane perpendicular to the axis of wave propagation.

Such orientations are called the wave polarization directions.[41]

In an ordinary beam of light, sunlight for instance, no direction of polarization is favored. When one direction takes precedence, we speak of polarized light.

A polarizer is a device whereby light is filtered through polarizing. What does not "pass" through the polarizer is reflected back and/or absorbed.

A monochromatic wave is periodic in time; its frequency, represented by the Greek letter *nu* ν, is the number of oscillations per unit of time. It can be measured in Hertz (1 Hz = 1 oscillation per second). Likewise, we can use the period T, i.e., the duration of an oscillation (expressed in seconds, for instance). The period is thus equal to the inverse of the frequency.

This wave travels at the constant speed of light **c** perpendicular to the polarization plane. It is therefore also periodic in space.

To give an intuitive image of this duality, one can say that a monochromatic wave is a periodic phenomenon in space when you visualize it at a given moment, and in time when you focus on a given point in space.

---

[41] More complex polarizations may take place but we consider here only vibrations in one direction.

The distance travelled during one oscillation is called the wavelength and is represented by the Greek letter *lambda* λ.

In short[42],

λ : wavelength of the electromagnetic wave,
c : speed of light ($\approx 3.10^{10}$ cm/s),
ν : frequency of the wave,
T : period of the wave,
and thus: $\lambda = \mathbf{cT} = \mathbf{c}/\nu$.

We see that the shorter the wavelength, the higher the frequency and vice versa.[43]

The evolution in time of the oscillation, that is its shape, is given by Maxwell's equations. Its maximum value is its amplitude A. The square of the amplitude A^2 measures the wave intensity, i.e., the energy density it transports.[44]

The electromagnetic radiation is thus one of the means by which energy is transported across space.

But what about our photons?

Quantum physics tells us that Maxwell's classical waves, and more specifically monochromatic waves that make up this radiation, consist of a great number of massless elementary particles travelling at light velocity: these are of course photons, the electromagnetic energy quanta.

In order to understand the very nature of these quanta, let's recall Einstein's conclusion concerning the photoelectric effect, i.e., the emission of electrons by a metal exposed to electromagnetic radiation. They are particularly enlightening, and foreshadow as early as 1905 the quantum revolution that will take place a few years later.

---

[42] For notations and symbols, see end of the book.

[43] The composition of the electromagnetic radiation is represented by the "electromagnetic spectrum," i.e., the classification of the waves according to their wavelength or frequency, with at one end gamma rays and at the other end radio waves, and in between, UV, visible light, IR, microwaves, and radar waves.

[44] The total energy density transported is actually the sum of its electric and magnetic components (neglected for the sake of simplification). These two waves oscillate in the same plane but in orthogonal directions.

In the photoelectric effect, one observes the following:

- if the frequency of the incoming beam of light is lower than a certain value, no electron is ejected, and the threshold value depends on the metal.
- For higher frequencies, electrons are ejected with a kinetic energy[45] proportional to the frequency of the incoming beam of light.

The principle of energy conservation implies that the energy of the incoming photon must thus be equal to the sum of the energy needed to eject an electron from the metal and the kinetic energy of the ejected electron. That is,

Energy of incoming photon = Ejection energy + Kinetic energy of the ejected electron

This result is best interpreted by positing that the energy transported by the incoming photon is proportional to its frequency; indeed, the threshold frequency determines the minimal frequency necessary to eject the electron, and the kinetic energy of the ejected electron depends directly on the frequency of the incoming photon.

This statement is symbolized by the following formula:

$\mathbf{E} = \mathbf{h}.\, \mathbf{\nu}$, where $\mathbf{h}$ is a constant quantity called Planck's constant. In other words,

Energy of the photon = constant x (frequency of the photon).[46]

Light behaves thus like a stream of photons whose energies are proportional to their frequencies.

We will deal later with this relation $\mathbf{E} = \mathbf{h}.\, \mathbf{\nu}$, presumably written out for the first time by Max Planck around 1899, recovered and correctly interpreted by Albert Einstein in the quantum framework in 1905, which earned both of them a Nobel Prize and rightfully bears the name of "Planck–Einstein relation."

We will see (Section VI.I) how Planck's constant $\mathbf{h}$, which gives the ratio between the energy of a photon and its frequency, turned out to be a leading actor in quantum physics.

---

[45] Kinetic energy of an object = increase in total energy of a moving object compared to its energy at rest. It depends on its mass and its velocity.

[46] Energy can be expressed in electronvolts (eV).

III.3.1. *Our stroll comes to an end*

The theories and events sketched in this episode summarize the essential state-of-the-art physics of the first half of the 20th century.

They form the basis that will allow this science to undergo a dazzling development over the course of the second half century.

We have introduced, on the one hand, general relativity, imperative in all situations where gravitation is dominant, e.g., in astrophysics and cosmology and, on the other hand, quantum physics, imperative in the atomic and subatomic world.

We will attempt to follow the developments of each of these important chapters of modern physics, in Episode V for the subatomic world and in Episode VI for cosmology, two fields which will meet surprisingly and intimately, two fields where François Englert played a crucial part.

Intermezzo: What is a Researcher?

An absurd question, some would say, for there are truly as many ways of being a researcher as there are researchers, just as there are as many ways of being a poet as there are poets.

They do share, nonetheless, a certain number of common features and it might be interesting to highlight the divergences as well.

Researchers are by definition people who are curious, not content to merely learn, but who want to find out for themselves an ever more pertinent answer to the questions they ask.

Whatever the field may be, before setting out to discover, they need to have acquired all the basics. For a physicist, that means an overview of the state of the art in physics, a mastery of the tools of mathematics, which is the language of physics. The university curriculum will normally ensure these fundamentals, provided that the institution resists the temptation of early specialization. This global trend to specialize is on the rise, in fact, and is liable to seriously jeopardize the quality of research, for maintaining a "generalist" notion of knowledge is always beneficial: countless examples demonstrate that an answer to a question arising in one field of science is often found in a different, sometimes remotely related field.

Universities should therefore be eager to provide every student with a broad scientific culture, a breeding ground where the future researcher will draw inspiration. Books, classes, labs, seminars, these are all classic modes of knowledge transmission, but they don't take into account the young rebels who will always prefer the road not taken. Those who, like Robert Brout and François Englert, are convinced that all knowledge is accessible by making it one's own, a personal path, and not exclusively through an academic curriculum. Hardly surprising, then, when it is often among these rebels, these impassioned enthusiasts, these unbridled men and women, that the most innovative researchers emerge. And the university has the duty to acknowledge and welcome them.

Faced with the enigma they have chosen to decipher, guided by their intuition, the moment's inspiration or the memory of analogous enigmas previously resolved, some researchers, those with pockets stuffed full of bits of knowledge, ancient, modern, long forgotten, retrieve one or the other, examine it thoroughly, lovingly, the way a luthier examines whether or not some piece of precious wood will one day produce music. Can this knowledge be applied somehow to untangle the enigma?

On the other hand, some prefer to approach an enigma without prior assumptions. Robert Brout was one of those. Unconcerned by the stranglehold of certain academic structures, he enjoyed attacking problems from scratch, as he used to say. There are an infinite number of scenarios, and researchers needn't "subscribe" to any single approach. In the particular case of François and Robert Brout, we shall see how two markedly different approaches can prove complementary, aligning and producing a particularly rich collaboration.

Though different kinds of approaches are available to a researcher, all of them call for imagination, intuition, and boldness: to imagine new modes of correlating observed phenomena, to devise original mechanisms, to cast doubt on well-established theories, and all of a sudden, among the jumble of images and concepts, to recognize the entry point that will enable the sought-after breakthrough and a foothold in a new paradigm.

Back in the 1960s, when François and Robert were grappling with the mechanism behind the mass of elementary particles (Section V.3), neither was a specialist in elementary particles or quantum field theory, which was hardly popular at the time, but both were convinced that a certain ingenuousness predisposed a person to creativity. And who could argue? After all, it was this first large-scale work that would earn them the Nobel Prize.

We often assume that science moves forward thanks to systematic deduction. This is not the case, according to François. Deductive reasoning does intervene at every step of the process, but the leap from one level to the next can only be achieved through a creative, transcendent act, a non-rational move, where the unconscious plays a role, and which the researcher will later reconstruct in a rational mode since most editorial committees of scientific journals are not yet willing to accept dream narratives, daydreams or otherwise, that give rise to beautiful equations put down on paper.

François often experienced at his very core this role of the unconscious/subconscious. On more than one occasion, he chanced to see the answer to this or that nagging question in the course of a dream, or to find his way there through the vicious maze of a sleepless night, or sense an inkling of a solution during a short nap. The omnipresence of a couch in every office space he ever occupied stands as testimony.

As François says, *New ideas are often buried in our unconscious world where our connections are untethered from the constraints of daily life. Intuition is a genuine act of creation, and it happens sometimes that we better perceive the world of fundamental research by listening to a Mozart opera than by attending a university lecture.*

As an echo to the words of the great mathematician Michael Atiyah: *During daylight hours, mathematicians verify their equations and their proofs, leaving not a stone unturned in their quest for rigor. But come nighttime and by the light of a full moon, they dream, floating among the stars and marveling at the miracle of the heavens. This is where they draw inspiration. Without dreams, there is no art, no mathematics, no life.*[47]

While still a young assistant at the ULB during the 1950s, every time some problem was nagging him, François would be overcome by the need to lie down, to escape from the contingencies of the moment; once immersed into that blissful state of detachment, he would feel a host of

---

[47] Both cited by Guy Duplat (*Particules de Vie* and a private conversation).

disparate notions rising out of his subconscious into his conscious mind: forgotten memories, scraps of knowledge, drafts of strange theories, dream fragments, bizarre images, all of which would gradually synthesize into an original vision of the question and lead him to perceive the relevant response. Only then would come the rational formulation, which was not always a simple matter, it goes without saying!

And so it was that as soon as he was granted his first university office, one priority was obvious: he had to find a couch. For this young penniless newlywed, there was but one way: the flea market. Esther and François found a not-too-nasty mattress. He set it right on the floor, covered it with an old spread, and it was ready to start its career as a "deliverer of creative neurons." Where the Freudian couch serves to mine the unconscious for the keys to better self-understanding and to assuage deep personal discontentment, the Englertian couch is a magical craft that laughs defiantly at the imperatives of rational deduction, guiding its navigator to the gates of the unknown that open onto an understanding that is often both acute and unexpected.

Episode IV

Robert Brout: A Lifelong Friendship

IV.1. François at the Dawn of the Sixties

In September 1959, François was awarded his doctoral degree in the sciences. With several papers under his belt, he had already laid the groundwork for his life as a physicist. It was then time to build the edifice. François realized he still had a lot to learn, in both physics and mathematics, but he then knew with certainty what he wanted: to understand the fundamental mechanisms of physics and to discover the definitive secrets of nature. The way forward had been charted. He had experienced at the deepest level the incomparable and insatiable pleasure of pitting the mathematically inspired constructions of his imagination against data emerging from contemporary physics. He had discovered the visionary power of his mind and understood that he possessed a formidable weapon.

He understood that in order to move forward, a fertile imagination was in fact indispensable. This was a time when familiar analogies so useful for arriving at an "intuitive perception" of theories in classical physics — physics at the human scale — no longer made sense in quantum physics, for example, the electron, the charged particle that conveys electricity, possesses a spin.[1] Spin is a quantum characteristic of a particle

---

[1] In quantum physics, spin is one of the internal properties of particles, on par with mass and electrical charge.

91

that can be assimilated to the kinetic moment of a sphere rotating on its axis, and which comes as such into play in the equations describing the behavior of an electron, except that the electron is not at all a sphere rotating on itself!

IV.1.1. *Physics: A comfort food for the soul*

The sensation of such certainty regarding his personal goals was something quite novel for François.

For the first time in his life, the vice grip that clamped on his heart since that day in 1942 when his life took a tragic turn was finally loosening. The former hidden child, separated from his family, had found a domain where he felt relieved of all his fears.

He also realized what a comfort to his parents his academic success had been and how they could rest assured that he was on his way to achieving great things in his career.

But despite the goodwill that had prevailed between François and his parents, communication remained as fleeting as always. Moreover, all his brain power and energy were being channeled into his enthusiasm as a young researcher.

It was only much later that he would understand the risks his parents took to save their children from the grip of the Nazis and, later, the sacrifices they willingly made to ensure that their children would seamlessly assimilate into their country of exile. When he realized how long it had taken him to reach this understanding, he experienced a sense of regret that nothing could erase.

IV.2. The United States

As we have seen, certain elements in François's publications reflected the concerns of an American physicist, Professor at Cornell University in Ithaca, New York, Robert Brout. In 1954, Brout was invited to ULB by Professor Prigogine, but François did not meet him on that occasion.

Later, when Robert Brout was looking to hire a research partner, preferably European researcher — born into a family of German Jewish émigrés, Brout was nursing a certain nostalgia for the old continent — he sought the advice of Pierre Aigrain, who provided a list of possible candidates, at the top of which were Pierre-Gilles de Gennes and François Englert, with whom he had stayed in close touch. Aigrain was said to have an unfailing knack for detecting talented researchers: de Gennes and Englert would both go on to win the Nobel Prize in 1991 and 2013, respectively.

It was François who would leave to join Robert Brout for a two-year contract at Cornell to work on "the study of solids."

To set foot on the American continent, he needed to get a visa.

One evening, François and Esther, his brother Marc, Robert De Mayer, and Georges Miedzianagora were invited to a party by one of Esther's friends; in this relatively bourgeois milieu, they came upon an American woman who turned out to be the Vice-Consul of the United States! There was much drinking, and Georges and Robert De Mayer were telling joke after joke, mostly frat-boy humor. The Vice-Consul was not amused. Yet, she was the person who granted visas, and when François later went to apply, the Vice-Consul issued a flat refusal. As it turned out, François's Soviet adventures, known to the authorities, were not unrelated to this denial, for during an interview at the Consulate, she questioned François quite aggressively about his stay in the USSR, about which she had undoubtedly been previously briefed by Interpol.

François contacted the agent from Belgian Security who once wanted to hire him as a double agent to discuss the problem.

They planned to meet in Porte Louise at the Café Flora, where François served up a narrative all his own: during a party at a friend's house, things didn't go well between him and the Vice-Consul, who was quite tipsy. He claimed she was putting the moves on him but that he declined her advances, all of which amounted to the real reason for her refusal to issue the visa. "Ah, women!" sighed the fellow, who then told him, "Look, it's a bit complicated, all that. What I need to know from you is whether or not you're a Communist. So, tell me straight, are you or aren't you?" To this, François replied, "I never discuss my political opinions, but I can tell you that I have never been a member of the Communist Party." Convinced, the man promised to help. What happened behind the scenes is unknown, but François did nevertheless obtain his visa, in the end!

Armed with the precious document, he left on his own for Ithaca in the fall of 1959. Esther and the two children, Michèle and Anne, who were still living on Rue de Tercoigne, would join him later.

Robert Brout was at the airport to meet him. François described the meeting this way:

> "Our first contact was incredibly warm: he came to pick me up at the airport in his old dilapidated Buick and took me to a café to have a drink. This first encounter lasted nearly the entire night. When we finally parted ways, we knew we would be fast friends."

Their friendship would indeed last a lifetime. And physics was not their only topic of conversation. Far from it! Brout, married and the father of two children, was, like François, curious about everything, had a passion for music and literature, like François, and had an Epicurean side despite the tormented soul that he managed to conceal.

As for "solid-state physics," which today is called "the physics of condensed matter," the two friends dove into it the very next day with gusto. They were interested in a variety of issues, in particular, ferromagnetic matter.[2]

Both were fearless scientists, unencumbered by dogmas and prescriptions customary in the academic world.

Robert and François were very different. Robert had a knack for translating into tangible gestures and images the most abstract concepts, where François, more Cartesian perhaps, was endowed with an intuition for the logical structures required for approaching a given problem.

They complemented each other to perfection, for each strove to understand how the other operated.

"Whenever we talked physics," said François, "each of us would feel a bit frustrated, in a way, at not being able to finish a sentence, because the other would step in, and gleefully so, apparently."

They soon wrote their first joint paper,[3] followed shortly thereafter by a second.[4]

The first article represented an outgrowth of François's doctoral dissertation, while the second was more formal and general.

The two friends came up with new approaches in the field of statistical mechanics, an area where Robert excelled,[5] while François found in quantum field theory,[6] somewhat neglected by physicists at that time, formal analogies with statistical physics that he took great delight in developing. They were studying magnetization of metals in particular.

---

[2] Ferromagnetic matter is spontaneously magnetic whenever the temperature falls below a certain level (the Curie point). In this context, we can visualize the spin of an electron as a small magnet. (Recall that a charge in motion creates a magnetic field). At low temperatures, the spins align and create magnetization. At high temperatures, thermic agitation prevents alignment. Above the Curie point, magnetization is nil. The matter is then in a paramagnetic state, having lost its spontaneous magnetism. The passage from the ferromagnetic phase to the paramagnetic phase is an example of phase transition.

[3] Brout R. and Englert Fr., "Dielectric Formulation of Quantum Statistics of Interacting Particles", *Physical Review*, 120, 1960, pp. 1085–1092.

[4] Brout R. and Englert Fr., "Linked Cluster Expansion in Quantum Statistics," *Physical Review*, 120, 1960, pp. 1519–1527.

[5] Statistical mechanics studies the evolution of physical systems consisting of a great number of items (atoms, molecules, ions, photons, and elementary particles).

[6] See Chapter V.2.

They would use their hands to better understand the spin configurations in a solid. Fingers pointing upward, fingers pointing downward, and when the spin can take all directions, imitating alternately the movement of waves, neatly depicting a spin wave, the spins influencing each other, like a propagating disturbance.

Robert had a hunch: since, in a ferromagnetic solid, magnetizing diminishes when the temperature rises, the spin wave frequency should tend toward zero at the time of phase transition. How could one express this property in the theory of magnetization?

The question sparked François's interest and followed him home that evening. The next day at sunrise, after a night of feverish labor, a solution was emerging. In a state of high elation, François scrawled out the equations. He had developed a whole theory, calling to the rescue a theorem of statistical mechanics that linked the correlations between spins in the absence of an exterior field to the response of spins in the presence of a magnetic field. He applied this theorem to spin waves. The curve of spontaneous magnetization took shape. He wrote down the equations, which proved valid! Everything aligned, and the results were more productive than even he anticipated.

This success filled François with intense joy and finally provided him with total self-confidence as a physicist. The next day, Robert Brout proved a harsher judge, which surprised François, since Hermann Haken, Visiting Professor at Cornell with whom Brout was also working on the magnetization theory, expressed only the highest praise. After dinner and lengthy discussions, Brout accepted François's solution. Later, he would explain his reluctance thus: *This is the beast in me...*

François's fine contribution would be published under his name alone in *Physical Review Letters,*[7] where he would pay tribute to Robert's intuition.

Robert Brout was a warm, generous, and quite complicated person. When he spoke, one sensed that he was watching himself talk, as if there were someone inside listening and judging him.

---

[7] Englert, F., "Theory of a Heisenberg Ferromagnet in the Random Phase Approximation." *Physical Review Letters*, 5, 1960, pp. 102–103.

Martine, Robert's wife, from Central Europe as well, was also a troubled soul. She met Robert in the US while hitchhiking. At times, she felt crushed by her husband's overbearing personality.

But despite their respective quirks, nothing could erode the friendship of these two physicists. Spending time with his philosopher friend Georges Miedzianagora had taught François how to find the paths that led to others. Despite his often scathing humor, François is full of tenderness for those he loves — tenderness, but also curiosity. He was forever intrigued, for instance, by how his friend's brain worked, this man who, when confronting a conundrum in physics, would immediately form a mental image of it, whereas he, François, would deal instead with decrypting the logical relations between observed phenomena. In order to better connect with his friend, he wanted to understand from the inside how such images arose, despite the fact that the objects of their research seemed precisely like the hardest thing to "imagine."

This question vexed him, until one night, there was a miracle! He had been stumped for hours on a problem of solid body structure, and finally just fell asleep. When he woke up, the solution was there, in his head, thanks to dream imagery! He was able to slowly piece his dream back together, seeing atoms, just as they were, taking up their positions in the crystalline solid and then reacting to stimuli of the outer field, moving faster with a rise in temperature, a hushed ballet in the dead of night, where he saw graphs and graphics gradually take shape. Pure joy! This kind of "illumination" would occur again, notably during his research linked to icosahedral viruses. Here, it was his engineering background that resurfaced: mechanics of materials, modes of vibration, instability, distortion, etc.

In April 1960, Esther placed Michèle and Anne with a "godmother" and went on to join François in the United States for a few months. In the lead-up to this visit, François unearthed a car selling for only $150, even more pathetic than Brout's old Buick. The couple had planned a cross-country road trip. But they scarcely made it to Chicago, their first stage, when some absentminded driver rammed into them: no injuries but extensive damage to the vehicle. Fortunately, François happened to have the address of some friends of his parents who lived in Chicago, and who immediately contacted a shrewd lawyer. The lawyer met up with François and Esther, along with an equally clever doctor, a neighbor of the friends. In under an hour, a "deal" was struck: the absentminded driver would pay $600 in reparations, $300 for the lawyer and $300 for François, who considered this a real windfall since this money would allow him to purchase a less pathetic automobile. They then headed south to St. Louis, along the Mississippi, to one of the birthplaces of the blues, jazz, and ragtime, and then on to Arizona, Nevada, and its salt flats, and drove back east along the northern route, through Wyoming and the wonderful Yellowstone Park.

That same year, when François was invited to a conference in Prague, he took the opportunity to spend his summer vacation in Belgium with his family, after which he returned to the United States. The family would then join him for Christmas, Esther, Michèle, Anne, and their youngest, Georges, born in December 1960. Cash-strapped as usual, François rented a modest house in a mostly African-American neighborhood. Right when François was on his way to pick up his family at the airport in his increasingly dilapidated car, an epic snowstorm hit the New York area. The jalopy managed somehow to make it through the 600-km round trip and brought everyone home safely. The children quickly adapted to their new lodgings. Little Michèle (six years old) was struck by how beautiful the neighborhood children were, and wished she could be black!

In the United States, François discovered a place where Jews played a rather special role. At Cornell, as elsewhere in the country, many physicists were Jewish, and those who weren't occasionally pretended to be, so significant was the "Mitteleuropa" atmosphere in academic settings. Unlike European Jews who had survived the Shoah, American Jews spoke openly about their Jewishness. In movies as in literature, Jews represented themselves unapologetically, and this didn't appear to bother anyone, at least on the face of it. For the first time, François had a vague sense of belonging to a certain diaspora. But he would never yield to any kind of clannishness, he who defines himself as a citizen of the world, and a free man above all.

François took part in a number of seminars at American universities. His reputation as a promising young researcher spread throughout the country. In 1961, he was invited to give a talk at the University of Chicago. There, he met someone who would make a lasting impression: the

Japanese physicist Yoichiro Nambu, eleven years his senior, who had just been appointed to the University of Chicago and would win the Nobel Prize in Physics in 2008 for his discovery of the mechanism of spontaneous broken symmetry in subatomic physics. At the dinner following the talk, François fell under the spell of this supremely intelligent and exquisitely courteous character. François would always feel deep respect for Nambu, whom he considered "THE Professor." He was also profoundly human — his students called him "the gentle genius" and "the seer." And like so many other Japanese, he practiced the art of never offending. Whenever a young researcher applied for a position in his department, he would never issue an outright "no." To reject a request, he would utter an unconvincingly yes. For an acceptance, he would declare "Oh yes!" with tremendous enthusiasm.

But in Ithaca, François's Soviet adventures continued to raise concerns with the administration. The American authorities seemed perfectly aware of his escapade in the land of the Soviets. McCarthyism had left its mark on American society, and to some, François, so overtly nonconformist, former participant in the Moscow Youth Festival, and Jewish, to top it off, must have appeared a bit suspect.

Once, at a party attended mostly by academics, François was flirting harmlessly with a young woman when her husband suddenly intervened aggressively:

"What do you do for a living?"

The question sparked François's instinct for provocation:

"I don't do anything, because I'm a Communist and the Russians pay me."

Two days later, he received a summons from the FBI.

It all went well, and François's high-risk pleasantry proved inconsequential.

This was also the time of the Algerian war for liberation, and François knew that several of his friends in Belgium were actively supporting the FLN;[8] it was the time of the so-called "suitcase carriers" who included

---

[8] National Liberation Front. Note that Belgians, sometimes risking their lives, helped Algerians during the war for independence.

Belgian Communists and progressives. But Europe was far away for François who was working tirelessly with Robert Brout. The two friends were grappling with numerous problems in the field of phase transition in statistical mechanics, notably ferromagnetism, superconductivity,[9] a highly complex phenomenon discovered in 1911 and elucidated in 1957,[10] and "spontaneous broken symmetry,"[11] which wasn't always really broken, but that's another story altogether.[12]

IV.3. The Return to Belgium

François's reputation as a brilliant researcher was spreading around Cornell and beyond. Appointed as Research Associate upon his arrival, he was soon made Assistant Professor. During the 1960–1961 academic year, as he was making plans to return to Belgium, he was offered the title of Full Professor, undoubtedly to retain him at Cornell, leapfrogging the Associate Professor stage, which rattled Associate Professor Brout a little. Despite his friendship with François, Brout felt apprehensive about the success of his "student," and feared that a rivalry was in the making. But he had nothing to fear, for competition mattered little to François, nor did he care about his friend's attitude: François wanted to go back home to Belgium. He received a call from Professor Géhéniau encouraging him to

---

[9] Superconductivity is a quantum state of matter. Certain materials become superconductors when they are cooled to extremely low temperatures, sometimes a few degrees above absolute zero. They then present two astonishing properties: they no longer resist the passage of electric current and they expel their magnetic flux fields, either by shifting them to the surface or by channeling them through flux tubes.

[10] Bardeen, J., Cooper, L.N., and Schrieffer, J.R. "Theory of Superconductivity," *Physical Review*, 108(5), 1957, pp. 1175–1204.

[11] A phase transition often comes with a spontaneous broken symmetry, for example, the passage from the paramagnetic phase to the ferromagnetic phase. Beyond the Curie temperature, the magnetism of a ferromagnetic material is nil; there is no overriding direction and the material presents rotation symmetry — it is isotropic. At low temperature, the material spontaneously acquires magnetization, which destroys the rotation symmetry, thereby creating spontaneous broken symmetry during cooling. Spontaneous broken symmetries gave rise to important discoveries in Field Theory, notably the work of Yoichiro Nambu.

[12] This is beyond the scope of this book.

return to ULB. Yes, but no longer as an assistant, that was a given. Professor Glansdorff, who valued François very highly, offered him a position of *chargé de cours,* roughly equivalent to Associate Professor. François accepted.

Just before he left, in fall 1961, François spiked a fever that would not subside. The doctor sent him to the hospital. François stayed there for nearly two weeks without the doctor's arriving at a diagnosis! Exasperated, François decided to leave for Belgium. His brother and a few friends came to meet him at the airport. Marc was worried sick, convinced that François must have cancer, perhaps even in its terminal stage, and booked him into the Institut Bordet! But in a few hours, Marc and his friend Yvon Kenis got to the bottom of this enigmatic fever: it was hepatitis, as proved by a simple blood test. François came away from the whole experience with a poor opinion of the American medical system.

Recovery would be long. For several months, François experienced extreme fatigue. The family rented a small apartment for a few months, until settling into a place on Avenue de l'Armée, in Etterbeek.

As soon as he had the strength to do so, he got back in touch with Robert Brout. The separation had been as hard for François as it had been for Robert. Could the close collaboration between two creative minds that had produced such remarkable results be somehow interrupted?

Of course not — it was unthinkable! Friendship would triumph and radically change the situation: a few months after François's departure, Brout decided to relinquish his chaired position at Cornell and follow François back to Belgium. He obtained a Guggenheim grant and got a position in the ULB "Pool de Physique." The Pool de Physique was in fact the result of a proactive initiative by Professor Glansdorff, who would become its director. The pool was to become an important body that oversaw the teaching of physics at all levels at the university. Paul Glansdorff was not only a highly competent director of scientific research but also a resourceful manager with a knack for obtaining subsidies from the administration and getting them to hire talented collaborators. Robert Brout joined the section where François Englert was working along with Harry Stern, Pierre Nicoletopoulos, and later Jean-Marie Frère, Claude Truffin, and Marie-Françoise Thiry. This section would become the Department of Theoretical Physics, which did not emanate from the institution but rather from an initiative by Jean-Marie Frère and François: a crafty little trick, if truth be told, consisting of a decision one fine morning to have some letterhead stationary made that read "Department of Theoretical Physics" which they would start using for all their academic correspondence! A virtual creation that became a reality.

Today, we would say that their move created a "buzz": academics the world over would henceforward deal solely with "The Department of Theoretical Physics."

Upon his arrival in Belgium, with Martine and the two children, hospitality was extended by Annette Deletaille (II.2), where François would join them for brief stays. The Brout family moved then into an apartment on Chaussée de La Hulpe, in the Uccle neighborhood of Brussels, where Robert organized a particularly memorable and alcohol-soaked party, American-style. Among the guests were Paul Glansdorff and his Pool de Physique colleagues.

Later, Brout had a house built in Linkebeek, where he could at last indulge one of his many passions: gardening.

Hardly a coincidence, then, that the Colloquium organized at ULB in honor of his 60th birthday was entitled "The Gardener of Eden."

In 1966, Robert Brout was granted Belgian nationality.

Robert was fascinated by craftsmen, and he himself was quite good with his hands. One day, he decided to have an armoire built by a carpenter on Rue Haute, in the Marolles neighborhood.[13] The highly skilled carpenter, however, was naturally lazy and could never work on his own. Not to worry, since Robert showed up every day to keep him company, working on his physics problems off to one side while monitoring with great delight his armoire as it was taking shape.

Upon his return to Belgium in 1961, François was made Associate Professor (*chargé de cours*) and by 1964, he had been promoted to Full Professor. Brout also taught physics and became Full Professor at ULB in 1968. The two friends continued their work on statistical mechanics and phase transition, in particular, on ferromagnetism and superconductivity. This research would result in multiple publications, adding to the body of work from François's American period.

---

[13] An old working-class neighborhood of Brussels, now a tourist destination.

In teaching as in research, François and Robert gave priority to exploring outside the box, gaining the reputation as being two unconventional professors. The following is what one of their former students, Jean-Marie Frère (who would succeed François in 1998 as head of the Theoretical Physics Department), said:

"It's worth recalling the boldness and enthusiasm back then. These young professors dared to innovate. Like when they introduced quantum mechanics in the second year, it was revolutionary. And they did it with style! François didn't hesitate to use the mathematical tools he needed, but without getting overly technical. He used Dirac's method (and manual), immediately presenting a profound conceptual vision, and also based his pedagogical approach on that of Feynman. A brilliant three-hour class (François is a convincing orator), and on a Saturday morning, no less!"

This originality in both teaching and research would not be appreciated by the institution. At ULB, from the perspective of the administrators, they came across as "wacky," even though they were increasingly esteemed elsewhere in the world.

For example, when François received the European Physical Society Prize, along with Robert Brout and Peter Higgs, it was the KU Leuven,[14] and not ULB, that would organize a surprise conference in honor of François and Robert followed by a surprise banquet! And never did his alma mater consider submitting their names as candidates for the Academy of Sciences.

François found that ULB's lack of interest worked to his advantage: he was left in blissful peace, free to do his work as he saw fit. But he felt they were being unfair toward Robert Brout, who had left a prestigious academic institution to come support cutting-edge research at a somewhat overcautious Belgian university.

Intermezzo: Robert Brout

It was during the 1965–1966 academic year that I first took a class with Robert Brout.

---

[14] Katholieke Universiteit Leuven, the oldest university in the country, founded in 1425.

All the students were impatiently looking forward to meeting this "Yankee," whose reputation as a goofy genius preceded him. A prof in jeans, sandals, and a hand-knit sweater was relatively rare at the time — but more so, a heavy American accent made worse by a slight stutter, an unkempt beard, all topped by an extraordinarily thick head of hair, and beneath those untamed curls, thick eyebrows, eyes sparkling with intelligence, a face both childlike and tormented that would suddenly light up with a mischievous smile, and especially, the astonishing way he was able to explain physics with his hands.

Unlike François, Brout didn't pace back and forth on the dais; after covering the blackboard with equations, sometimes verging on illegibility, he stood in place, attempting to explain what he had just written out, gesticulating broadly, "drawing" in the air these highly abstract notions. While still talking and gesticulating, he would begin erasing with his fist one or another equation, adding between the lines all kinds of mysterious symbols. His sweater would be covered in chalk, his hair dusty with it. Head cocked slightly sideways, his gaze feverish, Brout communicated his passion, stimulated our imagination, led us into a strange adventure where we would sometimes feel at sea, but never anxiously so, for we sensed that behind the professor and scientist, there was a friend who would never let us down.

And then there was Professor Englert, always ready to light our way. Never did I see François or Robert refuse to take the time to answer our many questions.

Episode V

The Genesis of the Brout–Englert–Higgs Mechanism

The next two episodes, fifth and sixth, tell the story of two fundamental discoveries owed to the work of François Englert, Robert Brout, and a few fellow travelers. For readers unfamiliar with the world of quantum physics, these episodes will probably be challenging simply because they will involve challenging notions, but since this is about François, for whom research is such an integral feature of his life, it seemed essential to present its most critical stages. We earnestly hope that uninitiated readers will fearlessly embark upon this journey into the unordinary and willingly go with the flow of the words without seeking too hard to unravel the mystery, for we know that by the time they land on the shores of this world that had seemed so impenetrable, they will have acquired a certain intuitive knowledge that will remain as the memory *d'un rêve étrange et pénétrant.* (Paul Verlaine)[1]

V.1. 1964 — A Visionary Paper

To fully appreciate the amazing events recounted below, it is essential to recall some facts and introduce some definitions.

About fundamental laws: during the first half of the 20th century, one might have thought that all observed physical phenomena were ruled by

---

[1] … of *a strange and penetrating dream.*

105

two fundamental laws: classical General Relativity which generalizes the Newtonian theory of gravitation and Quantum Electrodynamics (Section V.2) which is the transcription in quantum physics of the Maxwellian electromagnetic theory. Both theories are remarkably verified experimentally, and both deal with long-range interactions.

However,

The existence of subatomic structures implies the existence of short-range interactions, acting between objects at distances comparable to the size of atom nuclei.[2] Such interactions must be involved in building the nuclei, responsible for their cohesion and their eventual desintegration, but they must have negligible effects at the much larger scales we experience every day. However, nobody had yet succeeded in incorporating such short-range interactions in the realm of physics at that time.

V.1.1. *And what about elementary particles?*

We have met in Episode III electrons and photons. Electrons can be described in quantum physics by the Schrödinger equation (Section III.2). The photons are the quanta of light, defined as such by Einstein in 1905.

Two large families encompass all elementary particles: the Bosons and the Fermions.[3] The electron is a fermion, the photon is a boson.

— For fermions, a quantum state can be occupied only by a single particle of a given type. This implies, inter alia, that the different electrons "gravitating" around the nucleus of an atom belong each to a distinct quantum state and therefore do not all fall in the lowest energy state. This is what is called "the exclusion principle."[4]

— For bosons, a quantum state can be populated by an arbitrary number of identical particles. Waves filled with such bosons have a classical

---

[2] These subatomic distances are of the order of a thousandth of a billionth of a millimeter!

[3] The names fermions and bosons were chosen by Paul Dirac in 1959 to honor, respectively, the physicists Enrico Fermi and Satyendranath Bose.

[4] This property is crucial to understand the periodicity observed in the Mendeleev's Periodic Table (Section III.2).

limit, for instance, the electromagnetic monochromatic wave populated by a large number of photons of a given frequency.

All known matter is composed only of fermions and of bosons.

Among the bosons, there will be one, a rather peculiar one, which will forever be associated with the names of Robert Brout, François Englert, and Peter Higgs[5] since the announcement of the 2013 Nobel Prize in Physics, but which, for those knowledgeable in the history of Science, has been known under the acronym BEH-boson since 1964, when its existence and properties were predicted. We shall get to know it better in Section V.3.

V.1.2. *Let us go back to our two protagonists, François and Robert*

Between 1961 and 1964, they published many papers and the "Pool de Physique" acquired a strong reputation all over the world. They made contact with the "Service de Physique Mathématique" directed by Professor Jules Géhéniau, who studied elementary particle physics. The issue of working together was raised, but Géhéniau's group was reluctant to confront the determinedly innovating concepts driving the Pool physicists, a reticence Geheniau would later regret.

In 1964, François became Full Professor, and in June 1964 François and Robert sent for publication to the *Physical Review Letters* the three-page paper that would be made famous some half a century later by the 2013 Nobel Prize in Physics[6] for Englert and Higgs. Brout, who equally deserved the Prize, was not nominated because he had passed away two years before, in 2011, and according to the rules of the Nobel Foundation, the prize is never awarded posthumously. Their paper presented an original mechanism to explain how certain particles, a priori massless, acquire

---

[5] Peter Higgs is the British physicist who, quite independently, assumed the existence of the same scalar boson in a paper published a few weeks after the paper by François and Robert (Section V.3.).

[6] Englert, F. and Brout, R. "Broken Symmetry and the Mass of Gauge Vector Mesons," in *Physical Review Letters*, vol. 13, no. 9, 1964, pp. 321–323.

a mass, and proposes, as a follow-up, the use of this mechanism to generate short-range fundamental interactions. Of course, this is the mechanism with the predicted scalar boson as its basic constituent, which will be branded with the patronymic acronym BEH.

This paper was for François and Robert their very first publication in the domain of elementary particles.

In contrast with their earlier publications, which mainly dealt with the study in statistical mechanics (Section IV.2) of ferromagnetic materials and of their phase transitions, this paper presented the successful research of new fundamental interactions.

This is the occasion for a short digression about a very important aspect of research that is rarely emphasized and that played a significant role in François's approach to the problem.

Fundamental research tries to uncover the fundamental laws that govern observed phenomena. At first sight, it is different from the so-called "applied research" which aims to uncover "derived laws" resulting from the application of fundamental theories to particular materials (for instance, superconductors and semi-conductors). These derived laws often lead in a natural way to practical applications and to wide-ranging technological progress. But they also play an important role in fundamental research by virtue of the models and the new concepts they help to uncover. For instance, the general concept of spontaneous symmetry breaking was discovered while analyzing phase transition (Section IV.2) and was thereafter often applied to fundamental research.

Fundamental laws, derived laws, and the applications they inspire are intimately linked but it is not a one-way street, a "descending" direction from fundamental to applied. It is rather a reciprocal relation: fundamental and applied research feed each other. Applications not only enable the verification of the fundamental laws but are also a source of inspiration to discover suspected but still unknown extensions of the fundamental laws. François was quick to intuit this reciprocity and used it consistently.

To clarify the origin of short-range interactions, François and Robert had a feeling that they would have to explore the perspective of new fundamental laws.

In the early sixties, elaborating a consistent theory of short-range interactions seemed indeed to raise insuperable obstacles.

Two such interactions had been identified: the so-called "weak interactions" and "strong interactions".

In 1930, Fermi introduced the notion of weak interactions, responsible in particular for the radioactive disintegration of neutrons.

Around 1964, there arose the hypothesis of the existence of "quarks" which would be the constituents of protons and neutrons. Quarks were thought to be electrically charged fermions, which would interact not only with the long-range interactions of electromagnetism and short-range weak interactions but also with other short-range interactions, which will be labeled "strong interactions"; these were believed to be responsible for the cohesion[7] of the nucleons composed of quarks.

Thus, one imagined already at that time four types of interactions: gravitational, electromagnetic, strong, and weak — the first two being long range, and the last two short range, more difficult to analyze experimentally and not yet theoretically understood.

Before we unveil to the reader how an explanation finally emerged in 1964, we must introduce the tool which allowed François and Robert to come to terms with the enigma.

V.2. Relativistic Quantum Field Theory (QFT)

This title is liable to scare off uninitiated readers. But they should rest assured: Quantum Field Theory (QTF, implicit relativistic) for a long time scared off genuine physicists as well.

This theory, which developed throughout the 20th century, is the quantum generalization of the classical theory of fields. The quantization of Maxwell's Electromagnetic Field Theory, named "Quantum Electrodynamics," is one of its greatest achievements.

QFT is based on a highly complex mathematical arsenal, tricky to handle for the untrained, but it has gradually proved marvelously adapted to the description of elementary particles and their interactions. In fact, QFT generalizes and simplifies the interpretation of a great number of phenomena.

---

[7]The cohesion forces between nucleons are about one million times stronger than between atoms or molecules.

The scope of this book allows for only the briefest outlines of a few of its features.

Let us recall that in Episode III (Section III.2), we introduced the quantization of a massive non-relativistic particle (the electron) characterized by its position and its physical parameters (mass, electrical charge, etc.). Quantization was performed by associating to the particle a wave function whose evolution is determined by a "Schrödinger equation," interpreted as representing the probability of finding the particle at any point of the wave. This method is not directly applicable to relativistic particles,[8] and thus even less applicable to massless (relativistic) photons, and consequently, is ill-suited to quantize Maxwell's classical theory of electromagnetism.

Although one may try to generalize the Schrödinger equation to the relativistic domain, in QFT, a convenient general alternative procedure is available that has proven more useful in the study of elementary particles interactions. To achieve this, to each bosonic elementary particle (such as a photon) characterized by its momentum[9] $\mathbf{p}$ and its energy $\mathbf{E}$ one associates a particular classical field. In their QFT process of quantization, the particles become excitations[10] of the field. For instance, the quantization of Maxwell's classical electromagnetic theory (namely, quantum electrodynamics) results in describing the field excitations as photons of energy $\mathbf{E}$ and momentum p. Thereby are our photons promoted in QFT to essential actors of the quantization of the electromagnetic field.

As for fermionic elementary particles, e.g., the electrons, one cannot construct a classical field, because such a field can contain an arbitrary number of particles, thus violating the exclusion principle. However, in QFT, one may still describe quantum fields whose excitations of given energy and momentum obey the exclusion principle.

---

[8] Whose speeds are comparable to the speed of light.

[9] Recall that for a massive particle, the momentum $\mathbf{p} = \mathbf{m} \cdot \mathbf{v}$, where $\mathbf{m}$ is the mass in motion and $\mathbf{v}$ the speed. For a zero-mass particle, this relation becomes $\mathbf{p} = \mathbf{E}/\mathbf{c} = \mathbf{h} \cdot v/\mathbf{c}$. All the notations are explained in the table in the back matter and in Sections III.1 and III.2.

[10] Intuitively, one can imagine the waves in a quantified field as entities animated by vibrations. The greater the vibrations, the more excited the waves, and the richer in particles the field and vice versa.

QTF thus allows for a general formulation for both boson and fermion elementary particles.

The American physicist Richard Feynman[11] endowed QTF with a magic tool that allows for a simple, intuitive visual representation of all the possible physical processes that elementary particles can undergo. These diagrams that bear his name depict the propagation of each elementary particle by a line and their local interactions by vertices where incoming and outgoing particles meet. These interactions must preserve, in addition to the electric charges, energy and momentum. They may involve the creation and subsequent annihilation of particles (e.g., a line joining two vertices). For these particles, which exist only during the time of the interaction they transmit, this may imply a violation of the relation between the energy E, momentum p, and the rest mass m. Such particles are called virtual particles. Their role is illustrated in the following diagrams.

Each diagram translates the precise mathematical expression allowing the computation of the probability of the process depicted by the diagram.

As an example, the following are the Feynman diagrams representing the disintegration of a (free) neutron into a proton with the emission of one electron and one electronic antineutrino[12] by the weak interaction transmitted by a W^- boson (Section V.4). The W^- boson is here a virtual particle.

The first diagram represents a global process where the neutron and the proton are approximated as elementary objects. The second diagram details the elementary particle constituents (the quark contents of the nucleons): the three quarks (d u d) for the neutron and the three quarks (d u u) for the proton.

---

[11] Nobel Prize in Physics 1965 "for his fundamental contribution to quantum electrodynamics."

[12] Neutrinos and antineutrinos are electrically neutral fermions with very small mass. For the notions of antiparticles and quarks, see Section V.4: The Standard Model of Elementary particles. In Feynman's diagrams, antiparticles are represented with reversed time-arrow.

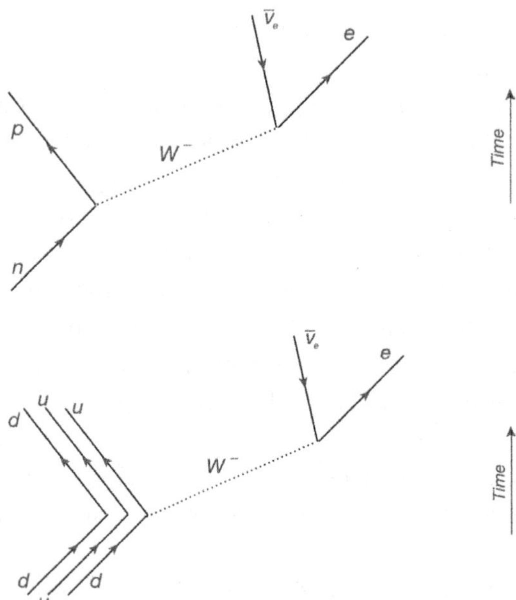

V.3. The Brout–Englert–Higgs (BEH) Mechanism

During the 1960s, many physicists were put off by the so-called Quantum Field Theory (QFT) approach to the study of (relativistic) fields in quantum physics. QFT had the reputation of being cumbersome and thus difficult to master. François was not aware of these considerations but he was very familiar with the formal analogy between QFT and the quantum statistical mechanics approach to phase transitions (such as ferromagnetism) which he had practiced a lot since he started working with Robert Brout at Cornell University.

He had the intuition that QFT should be just the tool to formulate a theory of short-range interactions, because it would make it possible to construct, at the level of atom nuclei, a new type of fundamental interaction which could be valid at the quantum level, a necessary concomitant for a predictive theory at the subatomic and subnuclear levels. And indeed the 1964 paper by François and Robert does not confine itself to the presentation of the BEH mechanism within its classical limits, a formulation equivalent to that presented shortly after by Peter Higgs, but does anticipate the study in all generality of its extension to the realm of quantum physics.

It was by analyzing the successes of the theories of long-range interactions, and particularly quantum electrodynamics (the quantum version of Maxwell's theory), that François and Robert were convinced that there should be a common origin to the fundamental short- and long-range interactions. They focused their attention on electromagnetism, guided by the already mentioned fact that its quantum extension in QFT (quantum electrodynamics) provides a successful description of electromagnetic phenomena not only at the atomic scale but at the scale of their nuclei which is 100,000 times smaller. This is remarkable because the quantum extension of a classical theory cannot always be performed (we shall actually see that the classical theory of gravitation, as expressed by Einstein's general relativity, still does not admit a quantum extension).

Impressed by this feature of electromagnetism, François and Robert, in order to approach the problem of short-range forces, got the idea to introduce Yang–Mills fields, i.e., classical fields similar to the electromagnetic field but whose quantum extensions do host "generalized photons."[13] These are new elementary particles which, like photons, are massless and generate as such long-range forces; but in addition they undergo point-like interactions among themselves. To cope with the existing short-range interactions, one could take advantage of these new interactions provided one finds a mechanism allowing these generalized photons to generate short-range forces. To this effect, some of these generalized photons have to acquire a mass, because in quantum theory, the range of the interaction transmitted by a (virtual) particle is inversely proportional to its mass. However, rendering massive generalized photons seems to contradict the very structure of the Yang–Mills fields! To circumvent this problem, François and Robert introduced then a new boson field whose quantum constituents are **scalar bosons**.[14] This is the birth of the BEH boson,

---

[13]Yang–Mills theories are a family of field theories introduced in the 1950s. Yang–Mills equations generalize Maxwell's equations, and generate quite naturally a generalization of Maxwell's photons. These theories have made possible a description of the fundamental interactions between elementary particles and form the basis of the Standard Model (Section V.4)

[14]A boson with spin = 0. This is a necessary condition to ensure the isotropy of the condensate under consideration. To discuss spin properties is outside the scope of this book.

another new particle that they assume to be massive and able to interact with any elementary particle, including generalized photons.

But how does this bold idea resolve the problem of the existence of short-range interactions?

Let us consider, as a thought experiment, a system consisting of a huge number of elementary particles at a very high temperature where all particles, with the possible exception of the BEH scalar bosons, would be of zero mass. Let us now lower the temperature; this "gas of particles" will then at some point undergo some kind of phase transition.[15]

Remember that, under ordinary atmospheric pressure, liquid water boils and becomes water vapor at 100°C, and inversely, that water vapor condenses to liquid water when the temperature decreases below this threshold.

Although the phase transition considered here is far more sophisticated than the liquid–vapor phase transition of water, this example provides an intuitive vision of the mechanism studied in this section.

Let's go back to the massless elementary particles at a high enough temperature, the scalar bosons coupled[16] with them, and the phase transition that this system will undergo when the temperature decreases and reaches some critical threshold: the scalar bosons will condense into a "liquid" phase which will fill the universe with a condensate (like a sea) of scalar bosons, of myriads of scalar bosons. Propagation within this condensate of a massless elementary particle coupled with this scalar condensate will result in a slowdown that can be interpreted as the acquisition of a mass by that very particle. This is the essence of the **BEH mechanism**.

Actually, what happens within the condensate is much more subtle than a simple deceleration: due to the elastic[17] collisions between the massless elementary particles and the scalar bosons they interact with, the particles will start "zigzagging" on their path; this will shorten the effective distance travelled by the particles, resulting in a decrease of their

---

[15] Reminder: in a general way, phase transitions link, at the macroscopic level, different physical states of a system whose constituents are the same at a small enough level.

[16] Reminder: coupling is a measure of interaction.

[17] That is, without energy dissipation

global velocity from its initial value which was the velocity of light. One may translate this measurable decreased global velocity as the acquisition of a mass by the particle, as only massless particles travel at the speed of light (Section III.1). We see that all massless elementary particles coupled to the BEH bosons acquire a mass through the BEH mechanism. In particular, the acquisition of mass by generalized photons transforms the long-range interactions they transmitted into short-range interactions.

We shall see in the next episode that the above thought experiment really did occur in the universe, as it appears to be confirmed by the experimental and theoretical developments of both cosmology and the theory of elementary particles: a beautiful illustration of the suggested merging of these two otherwise remote domains.

The BEH mechanism was presented in the mathematical language of the quantum theory of relativistic fields in its classical limit in the 1964 paper of François Englert and Robert Brout. Its full validity in quantum theory (its renormalizability)[18] was suggested on the basis of theoretical developments published in 1966[19] and presented by François at the 1967 Solvay Conference. The complete and impressive mathematical proof was given by Gerard 't Hooft and Martinus Veltman in 1971 using the Quantum Theory of Fields (Section V.5).

V.3.1. *François, Robert, and the 1967 Solvay Conference*

It was thus during the 1967 International Solvay Conference, dedicated to "Fundamental Problems in Particle Physics," that the foundations of the BEH Mechanism were discussed for the first time.

This is the opportunity to relate an event that sheds light on the relationship between Robert Brout and François Englert. Both were participating at the conference, simply as "auditors," meaning that they were not supposed to participate in the discussions. Since the BEH mechanism was their first work in elementary particle physics, they were thus hardly

---

[18]The standard term is "renormalizability." To be valid in quantum physics, a theory must be "renormalizable."

[19]Englert, Fr., Brout, R., and Thiry, M.-F. "Vector Mesons in the Presence of Broken Symmetry," in *Il Nuovo Cimento*, vol. 43, 1966, pp. 244–257.

known in this domain. However, when Steven Weinberg raised the question of the renormalizability of the mechanism, François could not resist expressing his conviction that the BEH mechanism was renormalizable, basing his arguments on the results obtained by his student, Marie-Françoise Thiry, in the doctoral thesis she was preparing under his direction. This statement provoked an animated discussion with Steven Weinberg and when the Conference secretary came asking François to write down his intervention, he refused, somewhat intimidated in front of an audience that he considered too competent in a domain where he himself was only a newcomer.

He was very surprised when, many years later, during an event in honor of the Dutch physicist Bernard de Wit[20] who had become a friend, Bernard asked him how he could, as early as 1967, defend the renormalizability of the BEH mechanism. François first did not understand how Bernard could be aware of his statement at the time, and was even more surprised when Bernard answered that his intervention had been published in the Proceedings of the Solvay Conference.

To François, this looked impossible as he had refused to write up his intervention. When he reported to Robert this conversation with Bernard, Robert told him how relevant he had found his intervention, and that, less timid than François, he had faithfully written it up and had presented it under the sole name of François Englert to the Editorial Committee of the Conference Proceedings. He did not, however, deem it necessary to let him know.

So, *the beast* could also behave gracefully, revealing a decency beneath it all.

Steven Weinberg was the first to apply the BEH mechanism theoretically. In 1967, he presented a unified treatment of quantum electrodynamics and short-range weak interactions[21] by coupling the BEH-bosons to the bosons that transmit weak interactions in the presence of electromagnetism. This unified theory of the so-called electroweak interactions, entirely based on the BEH mechanism, has been remarkably verified by experiments. For these results, Steven Weinberg was awarded the 1979 Nobel Prize that was shared with Abdus Salam and Sheldon Glashow.

---

[20] Former Ph.D. student of Martinus Veltman, like Gerard 't Hooft, and like Veltman, would become Professor at the Spinoza Instituut of the Utrecht University.

[21] Responsible in particular for radioactive disintegrations.

The mechanism imagined by François Englert and Robert Brout in 1964 was fully confirmed by the detection — half a century later — of the predicted scalar particle forming the condensed phase. This particle was identified in 2012 at CERN, after the replacement of the LEP collider (large electron positron collider) by the LHC (large hadron collider) capable of reaching higher energies. The properties of the detected boson particle confirm the BEH mechanism entirely (to the precision then reached). Its mass, not predicted by the theory, was found to be 125 GeV,[22] which is approximately 125 times the proton mass.

Many physicists paved the way that led to the discovery of the famous BEH boson at CERN. In addition to the precursors of the BEH mechanism, Nambu, Anderson, and Goldstone et al., there were those who worked and published during the half century following the papers of 1964 from Englert and Brout and from Higgs: Guralnik, Hagen, and Kibble in 1964,[23] Migdal and Polyakov in 1966,[24] then 't Hooft, Veltman, and Weinberg, and many others.

The BEH mechanism thus gave a mass to the bosons that transmitted the short-range weak interactions as well as to the fermions. By opening the way to short-range interactions, it led to a general theory of elementary particles, which during the second half of the 20th century would enable the identification of all known elementary particles, i.e., all constituents of the visible matter in the universe, and all the interactions governing them.

The BEH mechanism and its full validity in quantum physics (Section V.5), essential for reaching this achievement, are not just one more stage of progress in the domain of elementary particle. By extending to the nuclear and subnuclear levels the scientific understanding of the world, it constitutes a crucial step in our exploration of the universe on the path marked by Galileo, Newton, Maxwell, and Einstein. It proceeds from the

---

[22] GeV = Giga-electronvolt = a billion electronvolts. The equivalence mass-energy refers to the famous Einstein's equation $E = mc^2$.

[23] Guralnik, G., Hagen, C.R., and Kibble, T.W.B. "Global Conservation Laws and Massless Particles," in *Physical Review Letters*, vol. 13, no. 20, 1964, pp. 585–587.

[24] Migdal, A. and Polyakov, A. "Spontaneous Breakdown of Strong Interaction Symmetry and the Absence of Massless Particles," in *Journal of Experimental and Theoretical Physics* (URSS), vol. 51, 1966, pp. 135–146 (paper submitted in November 1965 and published in 1966).

oldest dream of humanity: to apprehend all questions raised by nature through a trustworthy and scientific approach. It is in this perspective that the mechanism and its scalar boson are the cornerstone of the Standard Model, the "official sum" of the presently acquired knowledge about elementary particle physics that will be presented in the next section (V.4).

For this remarkable contribution to the physics of the "infinitely small," François Englert and Peter Higgs were awarded the Nobel Prize in 2013.[25]

V.3.2. *A Problematic paternity*

At the time of its baptism, the BEH boson ran into a little problem. This offspring was claimed by three fathers, in fact. Peter Higgs, a British physicist, completely independent of the other two and right around the same time, had posited the same mechanism to explain the appearance of mass, but the precedence of Englert and Brout was indisputable, as François explained:

> *Our article appeared in the Physical Review Letters of 31 August 1964 where Higgs's article had only just been submitted.[26] Higgs, in fact, cites our work.[27] We therefore have antecedence. Which Peter Higgs readily acknowledges. Let us say that it was a co-discovery, in a way both independent but complementary. The mathematical approach was different. We were not acquainted. The particle in question came to be known as the "Higgs boson," and the name stuck, though competent scientists, for their part, know that it is the "Brout-Englert-Higgs boson." I actually prefer to call it something else, the "scalar boson," which better describes its nature.*

That said, we still find almost everywhere today, in both scientific publications and mainstream media, the label "Higgs boson" and oftentimes the "Higgs Mechanism." At the origin of this mix up is a simple error according

---

[25] Brout, R. Unfortunately deceased in 2011, was deprived of this consecration that he deserved as much as François Englert and Peter Higgs, but the Nobel Prize cannot be awarded posthumously.

[26] Englert, F. and Brout, R. "Broken Symmetry and the Mass of Gauge Vector Mesons," in *Physical Review Letters* vol. 13, no. 9, 1964, p. 321.

[27] Higgs, P. "Broken Symmetries and the Masses of Gauge Bosons," in *Physical Review Letters,* vol. 13, no. 16, 1964, p. 508.

to the "confession" made by the renowned physicist Steven Weinberg in May 2012 in an article entitled *The Crisis of Big Science* which appeared in the 10 May 2012 issue of *The New York Review of Books*, reviewing the book *The Infinity Puzzle* whose author, Frank Close, points out that Weinberg had acknowledged being responsible for the term "Higgs boson":

> *In my 1967 paper on the unification of weak and electromagnetic forces, I cited the 1964 work by Peter Higgs and two other sets of theorists. (…) As to my responsibility for the name "Higgs boson," because of a mistake in reading the dates on these three earlier papers, I thought that the earliest was the one by Higgs, so in my 1967 paper I cited Higgs first, and have done so since then. Other physicists apparently have followed my lead. But as Close points out, the earliest paper of the three I cited was actually the one by Robert Brout and François Englert. In extenuation of my mistake, I should note that Higgs and Brout and Englert did their work independently and at about the same time, as also did the third group (Gerald Guralnik, C.R. Hagen, and Tom Kibble). But the name "Higgs boson" seems to have stuck.*

Let us add a pinch of chauvinism so typical of greater nations, and we will better understand that nobody was in any great rush to set the record straight and give credit to "a couple of little Belgians."

In this regard, the Czech physicist Lubos Motl, former Harvard professor, who provides great insight into advances in physics in his blog "The Reference Frame," did not mince words in his dispatch on 2 July 2012, two days before the official announcement of the boson discovery, entitled *François Englert: a hero of the Higgs Mechanism*:

> *[…] François Englert of Brussels is one of the few (or two?) guys whose Nobel Prize should be pretty much guaranteed once the discovery of the God particle[28] becomes official...*

---

[28] Alas, yes, to the great dismay of numerous physicists, this boson was also nicknamed "the God particle." Originally, it was the title of a book, *The God Particle*, written in 1993 by the physicist Leon Lederman and the popularizer Dick Teresi. In fact, the original title should have been *The Goddam Particle*, because of the elusive, diabolical character of the boson, and was also an allusion to the parable of the Tower of Babel in Genesis (Tower of Babel − the CERN accelerator) but the publisher is said to have refused this title, too "shocking," in his view. God has nothing to do in the matter, in other words.

In 1964, before all the well-known papers written by Peter Higgs were released, he published a paper with Robert Brout — who unfortunately died in May 2011 — in which the foundations of the Higgs/BEH/ God mechanism were built.

[...]

Given these facts, you must find it somewhat incredible that while Google News counts 500+ hits for Peter Higgs in the recent 30 days, the same number for François Englert is 0 hits or 1 hit. It's kind of shocking. To say the least, it is another piece of evidence suggesting that a vast majority of the science journalists who are currently active are superficial slackers who just copy stuff from each other, including all the inaccuracies, distortions, intolerable oversimplifications, stupidities, and lies. This lousy culture is later imprinted into the whole population so as a species, we may perhaps boast that we're probably better than skunks but we still kind of suck and the inkspillers in the media are an important component of this fellatio.

[...]

I have no problem with using the word "Higgs boson" and even "Higgs mechanism" etc. It is catchy, efficient, and it has a true core. But if someone omitted François Englert from the hypothetical Nobel Prize that Peter Higgs will be getting, I would consider it a victory of the idiocratic, badly informed culture spread by the sloppy media and their undemanding consumers, culture that always prefers hype and group think over the truth. And that would be very bad because I do think that the physics Nobel Prize is one of the last awards that haven't been discredited by stunning blunders.

V.4. The Standard Model of Elementary Particles

*Our eyes do not perceive a fiery heat,
nor can they see the cold. As for voices,
we are not used to viewing them. But still,
all must consist of corporeal stuff,
since they can strike our senses, for unless
there is bodily substance, no object
can touch or itself be touched.*

Lucretius, *On the Nature of Things*

V.4.1. *A summary*

During the second half of the 20th century, the scientific community undertook the challenge of describing all observable matter in the universe, based upon the most recent findings in physics. In other words, they set out to classify all the elementary particles known to date, with all the laws that govern their behavior and their characteristics as confirmed through experimentation. We can now say, with a significant degree of certainty, that all visible objects[29] in the observable universe, from a drop of water to the human brain are made up of the building blocks that are fermions. These are divided into two large groups: quarks, which are the components of the nucleus, and leptons. Fermions feature a series of characteristics bearing poetic names. There are six flavors of leptons (electron e, muon μ, tau τ, and their associated v neutrinos), six flavors of quarks (up and down; charm and strange; top and bottom), and among quarks, three colors for each of the flavors.

For every elementary fermion, there exists its antiparticle, of the same mass and spin, but with opposite charge and "quantum numbers."

All these fermions are subjected to forces transmitted by "mediators" or "messengers," which are also elementary particles, bosons, which are all, with the possible exception of the hypothetical gravitons transmitters of gravitational interactions, excitations of "generalized photon" fields (i.e., Yang–Mills fields) as predicted by Englert and Brout (Section V.3). All fermions are subjected to four types of interactions transmitted by bosons:

- Photons are the messengers of electromagnetic interactions.
- Gluons are messengers of strong interactions that involve only quarks.
- W^+, W^-, and Z^0 bosons are messengers of weak interactions.
- Hypothetical gravitons are thought to be the messengers of gravitational interactions.

---

[29] We make the distinction between visible and observable, since all observable objects, i.e., detectable through experience, are not necessarily visible. The only objects visible are those that interact with the electromagnetic field. Dark matter, for instance, is observable but not visible (see Epilogue).

Note that the BEH scalar boson is not a messenger, but confers mass upon all the massive particles of the Standard Model.

Each elementary boson also possesses its antiparticle, but in certain cases, particle and antiparticle are identical (e.g., for photons).

The hypothetical graviton is thought to be the quantum of the gravitational field, just as the photon is the quantum of the electromagnetic field. In the same way that electromagnetic waves contain a large number of photons, gravitational waves are believed to contain a large number of gravitons. But unlike photons, they have not yet been detected individually because their coupling with elementary particles is extraordinarily weak. The gravitational effects are generally negligible in the analysis of the universe's constituents. This issue will be addressed in Episode VI.

All the elementary particles of the Standard Model are the ultimate components (to date) of all the known objects and of the interactions they undergo on Earth or elsewhere in the universe, be they living or inanimate.

To this concept of a "Lego universe," François Englert and Robert Brout provided the basic bricks whose pertinence has been confirmed by experimental findings.

L'école buissonnière: A Few Memories from May 1968

During the years subsequent to the publication of their 1964 article, François and Robert were caught up in the maelstrom of discussions about their work. Seminars and colloquia followed in fast succession. In 1967, at the Solvay Conference whose theme was "Fundamental Problems in Particle Physics," there were heated discussions about attempts to unify electromagnetic forces and weak interactions, and of course, about Englert, Brout, and Higgs's prophetic hypothesis. Sessions were lively and impassioned, resulting at times in real clashes.

But neither François nor Robert were scientists shut up in their ivory tower. Mindful of the sound and fury of the world, they were fully aware of what was happening around them, fast-paced events both troubling and encouraging. Our neighbor, France, would soon be capturing everyone's attention.

1968: the Golden Sixties were drawing to a close. Everyone recalls those years as a youth revolution, the cradle of a broader questioning of the social order that would result in movements that can be defined as insurrectional, all against the backdrop — and this is no coincidence — of a particularly unsettled political context: the Vietnam War, the Greek junta, the assassination of Martin Luther King, the Prague Spring, etc.

May 1968 would remain in the memory of all those who "were in their twenties in 1968," twenties, or a little older, for this wave of protest, remembered among other things for its poetic revolutionary slogans, swept through a large part of the world that year leaving no one unmoved. Although the initiators of the movement were largely high schoolers and college students, they were joined by young artists and workers as well, and eventually all sectors of society were affected. Some were enthusiastic backers from the get-go, while others panicked at this sudden and arrogant intrusion by youth culture into their adult world, while still others observed all this commotion with a certain disgust and unmitigated contempt. But among those who experienced first-hand what these young people were up to, it was teachers and academics who dealt with them on a daily basis who most quickly understood the importance of this movement and joined their ranks.

For our immediate neighbors in France, protests had been shaking up universities since the beginning of the year. The 22 March Movement, as it came to be known, started on that day when students occupied the administrative building at the University of Nanterre as a protest against the arrest of students carried out during an anti-Vietnam War demonstration. This movement, led mostly by student anarchists, Trotskyists, and Situationists, labeled in the media as *enragés* (mad dogs), was highly critical of consumer society, imperialism, and "your daddy's university." The movement also promoted total sexual freedom and called for coed dorms and the free circulation of male and female students throughout all university housing accommodations.

The closing of the Nanterre campus, decided on 2 May after a turbulent day-long meeting focused on anti-imperialism, triggered a shift of the movement toward the *Quartier Latin* and the Sorbonne. There, a multifaceted form of protest started taking shape, a questioning of all forms of authority, and that was where May 68 got off the ground.

On 3 May, the Sorbonne was occupied by protesting students, then brutally evacuated by law enforcement. The throngs of demonstrators was growing, paving stones were being pulled up and used to construct barricades during clashes with the riot police (CRS). The scope of the movement spread beyond all expectations and aroused sympathy among public opinion that expressed horror at the behavior of the CRS.

In Brussels, and at the ULB in particular, everyone was closely watching events in Paris. Wide-ranging protests against bourgeois university structures were developing, though the tone was far less explosive. During numerous Student General Assemblies (AG), the events of Paris were discussed, reported in detail by student delegates sent on information-gathering missions to the *Quartier Latin* in Paris. Action plans started emerging. The AG convened on 6 May in the big meeting hall of the university residence foreshadowed the daily Free Assemblies that would soon become the catalyst of university protest.

Alongside this student mobilization, an impetus driven by a core of intellectuals and artists, some of whom will soon be viewed by the young protesters as mentors, was to play a crucial role. It is worth mentioning the little tribe around Robert De Meyer, that longtime hotbed of protest and provocation.

Though Georges Miedzianagora and François were present from the very first Free Assembly onward, and would be a part of those professors and students who wanted to "kick the Administrative Board out of the University," the two friends were envisioning something on a very different scale.

For Georges Miedzianagora as for François, Belgium had stayed much too calm and the student protests too gentle. It was time to shake things up a bit.

François remembers:
One fine day in May, Georges Miedzianagora showed up at his place. He had been closely monitoring the situation in Paris. French Prime Minister Georges Pompidou called for a de-escalation, on 11 and 12 May, but Miedzianagora wished to see the movement radicalize. They had to further empower the revolt, to move it beyond national boundaries, to bring people together, from here and elsewhere, to fan the flames that will

rejuvenate the old world. Jérôme also stopped by François's place, though he was completely dejected at the prospect of a "de-escalation" in Paris, and threw in the towel.

Georges announced unequivocally,

"We are going to make a revolution!"

But before all else, the connection had to be made with the other protest movements going on in Europe.

The telephone would prove their principle weapon in this struggle.

They alerted a few friends who had just returned from Paris, among them Willy D. That evening, a mini "war room" came together at François's, at 6 Rue General Patton.

"We have to do something," François urged.

They decided to make phone calls to the "masterminds" of the other main student movements in neighboring countries — France, Germany, Italy, Great Britain, and the Netherlands — inviting them to come speak at the ULB. And they all came! Solidarity still had meaning.

The next day, 13 May, a gathering took place at François's: a representative of the German SDS,[30] a few of Daniel Cohn-Bendit's brothers in arms, and Italian, English, and Dutch student leaders. Discussions ignited: revolution, yes, but then what? Everyone pitched in suggestions.

François and Georges, on the other hand, remained very "Zen," leaving the "revolutionary leaders" to agitate, and left to grab a bite to eat in town. They weren't the militant type — while they did take delight in the anarchist, provocateur antics, they refrained from joining any programmatic revolutionary activities.

On the evening of 13 May, after a speech made by Melina Mercouri about the ruling Greek junta in the Paul-Émile Janson Auditorium, a large group made up of students and a few professors decided to stay in place and "occupy" the auditorium, taking their turn to express their exasperation with bourgeois education standards and structures. The students demanded that they be given a voice in how the university was run: it was time that the university's Administrative Board, composed of notables from outside the academic world, yielded power to a democratic,

---

[30] Sozialistischer Deutscher Studentenbund.

representative body. At this gathering, which would go on until 3 a.m., the young doctor Mony Elkaïm laid out the principles drawn up by the Paris *Assemblées Libres* (Free Assemblies). As the meeting broke up, the participants, who called themselves "the May 13 Movement," with reference to the "March 22 Movement," decided to reconvene the next day at noon at the ULB student residence. The following day at noon, students, professors, and staff took part in the Free Assembly held in the foyer of the residence. The "foreign delegates" invited by Georges, François, and Willy took the floor, lending the "Belgian" protest a new legitimacy, with the ULB insurgents sensing that they resonated with the events that were stirring up "old Europe."

Another free assembly was improvised in the afternoon in the main meeting hall of the residence. It was better attended, more protesters had been recruited. Things were heating up, people wanted to take action. François and Georges took the floor: enough discussion, they say. How about going to have a word with the Administrative Board?

After some reluctance expressed by a few, François came out with "Alright then, let's do it!"

Lots of remarkable things happened in this merry month of May. This is not the place for an exhaustive chronology. Books, memoires, and dissertations have recounted the events, oftentimes with approximate objectivity.

The following is a random sample:

– The very lively Free Assemblies that took place in the great hall of the student residence where various trends of protest were expressed, from libertarian internationalists to reformists of all stripes.
– The invasion of the women's residences by students demanding free access to all student accommodation by both men and women.
– The vote of a motion by 174 professors ceasing to acknowledge the authority of the Administrative Board and demanding that the board be replaced by a body elected by the whole university community — a historic moment of the ULB, for the previous uprising by the faculty dated back to the 1890s!
– And all the while, the university doing its best to crush the movement.

- The large-scale Free Assembly held in Janson Auditorium, in the presence of TV cameras, press photographers, and security personnel composed of ULB technical staff and non-academic workers.
- The Storming of the Administrative Board where the insurgents took pleasure in watching the administrators make a hasty exit from the premises before they all took their seats around the large boardroom table to discuss what was to come next. (People said that the "ex-administrators" decided henceforward to hold their meetings at the Villa Lorraine, one of the best restaurants in Brussels.)
- And finally, what was undoubtedly the most memorable event, a true venture for many: the occupation of the university, starting with the rector's office and other administrative quarters, places considered until then sacrosanct.

Despite the libertarian slogans that had started to tag the walls, the occupation took place in a quite rational manner and the seized premises would become the center of intense activity. Mimeograph machines were working overtime, tracts were distributed from one end of campus to the other. Communicating with the media was quickly proving essential, as the Free Assembly needed to make its positions known outside the ULB walls: a phone center was set up where students took turn fielding calls, a radio station broadcast 24 hours a day, and journalists poured in looking for updates. A Free Assembly was held daily in the *Salle des Marbres*, the Marble Hall, beneath the tutelary silhouette of a bronze god, a statue of Zeus, or was it Poseidon? — no matter, attendees discussed, voted, and remade the world.

Many artists came to join the insurgents occupying the ULB. Among them were the painters Arié Mandelbaum, Georges Meurant, Annette Deletaille, Roger Somville, and sculptor Joseph Henrion.

The walls were soon covered with drawings, comics, Chinese *dazi-bao*-style wall posters, and huge canvases illustrating such emblematic figures as Che Guevara, Mao, Marx, Lenin, Dubcek, Mandela, Lumumba, Angela Davis, Gandhi, Martin Luther King, Malcom X, and Louise Michel — Zeus/Poseidon was wrapped in banners, his extended arms providing the perfect hooks to hang their banderols.

The artists Somville, Meurant, and Mandelbaum painted broad-brushed frescoes full of revolutionary fervor, poets declaimed odes to the protest, and musicians held recitals. A whole little world was taking shape around the occupation. The daily Free Assembly expanded to include a parade of theorists of all persuasions, philosophers, artists, self-proclaimed revolutionary leaders, professors, trade unionists, anarchists, young militant Communists, Trotskyists, Maoists, and also militant workers, for the junction between student movement and workers movement had been established, albeit hesitantly. A handful of students had headed out to the factories to preach the liberation gospel, urging workers to participate in their free assemblies. Some tried their hand at agitprop: young actors recited Mayakovski, Nazim Hikmet, and Aragon standing in front of blast furnaces, not making many converts, though they were warmly welcome.

However, when the insurgents ventured into the city to distribute tracts, they were met with a less than enthusiastic greeting. This was when François and a smattering of students were confronted by a brutal charge of mounted police, swords drawn, in the middle of Brussels!

But keeping a clear head and feet on the ground, his main concern being the fate of the students, François got the Free Assembly to vote to hold a third session of exams. This request, which he would go on to defend before the Chancellor Marcel Homès, was finally granted, allowing a number of students, who while they had been "changing the world" had skipped class for over two months, sit for their finals and avoid flunking out.

But after a while, the occupation of the buildings started causing much vexation among the authorities and great concern in the political sphere. Security forces intervened on several occasions, sometimes heavy-handedly, sometimes more peacefully. The "ringleaders" were taken away in police vans, reappearing the next day looking not too banged up. Belgium would not experience the same level of brutality as did France. But the police presence naturally felt like an unacceptable provocation. Faced with the urgent need to defend themselves, the insurgents resorted to some rather original strategies, for although their backers among certain political parties and trade unions could provide a few very effective "bruisers," there simply weren't enough of them. So, the question was where to find some heavies who knew how to fight and who wouldn't back down when confronted by the police. Georges Miedzianagora, who meanwhile had become a mastermind of May 68, François, Ben Tursch, and Jean-Claude Garot put their heads together and proposed that they meet up with the "fascist students" at the *Gueule de Bois* ("hangover") tavern.

Despite earlier skirmishes that quite violently pitted the protesters and the fascists, the latter accepted a truce and promised, with no qualms, to come over and lend a hand. What interested them, in the end, was the pleasure of a rumble. The Flemish students, who had also occupied the university, took the same approach: they called on the VMO "tough guys," the *Vlaams Militanten Orde*! This unlikely cooperation between the far right and the 68er leftists gave rise to some charmingly surrealist scenes that were etched into the collective memory. For instance, when hundreds of police officers suddenly lined up along Avenue Roosevelt, facing off with hundreds of students chanting revolutionary refrains, supported by a fascist militia shouting absurd slogans. Or when the intrepid Ben Tursch dismantled Molotov cocktails that had been prepared together by protesters and overexcited fascists who were not quite sure where they were supposed to throw them. Or when the painter Annette Deletaille made bread-and-jam confections to restore a young fascist who had somehow been beat up by some leftist or other.

It was these kinds of uncanny scenes, among so many others, that provided the weft of the memories that so many would recall from the month of May 1968. Those nights on campus, the constant coming and going of students, discussing, joking, having a mug of beer together, music streaming from somewhere between the physics labs and the philosophy building, couples embracing on the grassy quad, the hubbub of free assemblies with their emotional highs and exhausted lows, their searing intensity, their heated debates punctuated by laughter and bursts of craziness, the walls covered in slogans, the delegations of students from other universities received with open arms, the solidarity, the brotherly dialogue with professors...

Such scenes would forever inhabit our memories, though the relatively disappointing outcome of May 68 was quickly forgotten. True, many things changed at the university, notably the status of the Administrative Board and of the staff, or a rethinking of the nomination process. But the machine's authoritarian character quickly recovered, dressed up as a "co-management" that never really worked very well. The professors were invested with new powers, not always with positive repercussions, and the students, dispossessed of their libertarian dreams, essentially lost interest in the issue. Increasingly depoliticized generations have followed the "May 68" generation. The protesters who found the beach beneath the paving stones[31] were replaced by the "who cares" generation and the triumph of individualism.

There were also a few victims among the assistant professors who had embraced the students' cause. Georges Miedzianagora was banned from teaching, and went on to found, with a few faithful companions, the *École Élisée Reclus*, recommencing an initiative taken in 1894 by a handful of progressive professors at the ULB when the university opposed the arrival of anarchist geographer Élisée Reclus who had, paradoxically, been invited by the rector. The École held classes for a time within the ULB, then moved to the atelier of the painter Georges Meurant. François, along with Ilya Prigogine and René Thom, taught at the *École Élisée Reclus*. But this courageous initiative was short-lived, for lack of funds.

Georges the philosopher published during this period a small work on quantum mechanics that left physicist François somewhat perplexed. The philosopher ended up moving into an old refurbished windmill in Spain, ever the dissenting, miscreant stoic, a kind of Diogenes for our times.

As for François, in the end greatly disappointed with the outcome of the movement that had held so much promise, a breath of freedom, he had no illusions when it came to the supposedly more democratic university that emerged from May 68. Disgusted at the new nomination process for professors which, instead of featuring more transparency, opened the way to the settling of scores among peers, he went back to his work, which he

---

[31] *Sous les pavés la plage* was an oft-repeated slogan in Paris during the May 68 insurrection, a reference to the paving stones pulled up to build barricades and to the sand beneath, the dreamy beaches of their imagined utopia.

had in fact never really interrupted — a dozen scientific papers managed to get written during the year 1968!

V.5. No Doubt, It is Renormalizable!

Gerard 't Hooft and Martinus Veltman

In 1971, the young Dutch physicist Gerardus (Gerard) 't Hooft, in his Ph.D. thesis, presented at the University of Utrecht, proved the validity of the BEH mechanism in Quantum Field Theory, and thus in technical terms its renormalizability.

His thesis director was Professor Martinus (Tini) Veltman, a physicist who knew QFT better than anyone and all the experimental as well as the theoretical details of elementary particle physics. The results of their collaboration were of capital importance. They provided the Standard Model, still in progress at the time, with a reliable mathematical formulation and theoretical tools which would allow successful predictions of the properties of already known or still to be discovered particles. This model, which aimed to describe all known elementary particles and their properties at scales even smaller than atomic nuclei, required essentially the quantum validity of the BEH mechanism and that was indeed what the work of Gerard 't Hooft and Tini Veltman had achieved.

For their remarkable "elucidation of the quantum structure of weak interactions," they would jointly receive the Nobel Prize in Physics in 1999.

During that period, François was working on the interpretation of strong interactions: was there some kind of duality between the weak and the strong short-range interactions?

While the BEH mechanism suggested the existence of a phase transition analogous to superconductivity (Section IV.2), François and Paul Windey showed that strong interactions presented an analogy with a "dual" phenomenon to superconductivity, where the electric charges of the superconductor would be replaced by "magnetic charges." This would be the starting point of their general theory of magnetic monopoles in QFT, generalizing the theory introduced by Gerard 't Hooft.

Gerard 't Hooft would become one of the most prominent physicists of his time, able to uncover and solve in an original way key problems of contemporary physics.

Tini Veltman liked to focus on experimental results and felt close to the visual physical perception of Robert whom he greatly appreciated. Impressed by the suggestion of the renormalizability of the BEH mechanism presented by François at the 1967 Solvay Conference, he would mention it later in a review of the subject:

> *[...] about the work of Englert and Brout, which slightly predates the work of Higgs. These are evidently independent pieces of work. Actually Englert and Brout saw clearly the connection with renormalizability, and they advised Weinberg of that, at the Solvay Conference of 1967. Students of the history of the Standard Model may want to check this little-known reference.*[32]

Tini Veltman, Gerard 't Hooft, François, and Robert would often get together and their friendship was always cordial and warm-hearted. François was fascinated by Gerard's insightfulness and appreciated how thoughtful he was toward his family, as he witnessed on the many occasions of their get-togethers.

In 2006, François and Mira were invited to a colloquium organized to mark Gerard's sixtieth birthday, a gathering attended among others by Lenny Susskind and Andreï Linde, both professors at Stanford. It took place on the lovely Frisian island of Vlieland. The University of Utrecht pulled out all the stops: each participant received a necktie designed and signed by Gerard, featuring all the elementary particles of the Standard Model, in color, except for the scalar boson, all black, since it was still hypothetical! François would wear this tie so often that it was almost threadbare. When François asked for a new one, Gerard granted the request, provided that François came to give a lecture at Utrecht. Deal struck!

As for Tini Veltman, he was a man who, in a heated discussion, was capable of indulging in some rather provocative statements. For instance, although not an anti-Semite, he jokingly told François and Mira the

---

[32]Veltman, M.J.G. "The path to renormalizability." Invited talk at the 3D Internastional Symposium on the History of Particle Physics, Stanford, CA, June 1992. In *The Rise of the Standard Model, Stanford Conference Proceedings,* pp. 145–178.

anecdote about being so warmly welcomed by Jewish American physicists because they of course thought that Veltman was a Jewish name. François didn't fall for it, and swore to himself that he would turn the tables on him when the occasion arises. One evening when François and Tini were having dinner at the home of a colleague and friend, Raymond Gastmans from the KULeuven, Tini launched into one of his wisecracks:

"Are you born in June? All intelligent people are born in June. I was born in June."

François retorted instantly:

"I don't need to be born in June because I'm a Jew!"

In the end, everyone had a good laugh.

Episode VI

On the Threshold of Cosmology

Voie lactée ô sœur lumineuse
Des blancs ruisseaux de Chanaan
Et des corps blancs des amoureuses
Nageurs morts suivrons nous d'ahan
Ton cours vers d'autres nébuleuses[1]

Apollinaire *(La Chanson du Mal-Aimé)*

VI.1. When "the Infinitely Small" Merges with "the Infinitely Large"

We should always bear in mind that the theoretical constructions that we just discussed, born from the imagination of physicists using the resources of the most subtle mathematics, have been verified by experimental results. To date, they have been confirmed both by astrophysical observations and by particle accelerators. In addition, they enable us to predict the existence of new interactions and new particles, as exemplified by the BEH boson.

---

[1] Milky Way O luminous sister/The white streams of Canaan / And the white bodies of women who loved/Dead swimmers we shall follow, gasping/Your course toward other nebulae.

We have seen that, in the second half of the 20th century, there developed close ties between the physics of "the infinitely small" and the physics of the "infinitely large."

VI.1.1. *But what do we mean by "infinitely small" and "infinitely large"[2]?*

In everyday life, we all have notions of what is small, very small, or large, very large. We easily evaluate in meters, kilometers, in grams, milligrams, tons, in seconds, in years, in centuries, sometimes in millennia …

When we use the expression "infinitely small," we evoke things that are really very small, for instance, invisible to the naked eye or even with a microscope.

When we say "infinitely large," we immediately have in mind the size of the universe.

Before the discovery of atoms, the size and other characteristics of the objects studied in physics were not essentially different from those encountered in daily life.

Likewise, before the discoveries by Einstein, first of special relativity and then of general relativity, the notions of space and time did not seem to require any clarification.

In the 20th century, something absolutely fascinating happened: the notions of space and time were endowed with complex properties which would lead to the unexpected consequence that the use of these very notions, which had appeared so natural to everybody, may have a limited validity.

More precisely: we did not imagine that we would not be able to continuously scale down length and time, so as to approach vanishing points of time and distance intervals as closely as we would like to, because there are values below which the physical reality will most probably become uninterpretable in terms of space and time and that this feature appears to be linked to the stranded quantization of general relativity.

---

[2]We use quotes on purpose, for various reasons, but also to indicate that these terms have a specific mathematical signification, irrelevant in the present context.

VI.1.2. *The limits of classical general relativity*

We have seen in Section III.2 that the classical theory of electromagnetism did not succeed in explaining physics at the atomic scale. At that level, quantum corrections become dominant with respect to the predictions of classical physics and thus require the quantum version of Maxwell's electromagnetism: quantum electrodynamics. This is all the more true at the nuclear level that deals with distances at least 100,000 times smaller. At such scales, one must take into account, in addition to quantum electrodynamics, the quantum weak and strong interactions of the Standard Model.

What about general relativity? We know that in general relativity, gravitational interactions are viewed as distortions of space and time. Henceforth, could there be limits to distances and time intervals below which these notions, commonly used in classical physics, would be blurred to leave room for new concepts revealed by the quantization of general relativity? This is a very daring question to ask as there is presently no reliable theory for the quantization of general relativity. Amazingly, however, we shall discover that it is nevertheless possible to formulate a reasonable hypothesis about the existence of such limits and even about their values.

Let us look more closely and reexamine Max Planck's daring hypothesis laid down in 1900[3] (Section III.2) of a quantization of the radiation emitted by the walls of the black body. Planck computed the density of this radiation (which depends only on temperature) by assuming that it resulted from the emission of "energy quanta" $h \cdot \nu$, where ν is the frequency of the emitted radiation and h a constant that Planck evaluated by comparing his theoretical results to the experimental data. This result, which concluded dozens of unsuccessful attempts, is still remarkably valid today. A few years later, the black body radiation as described by Planck would be interpreted by Einstein as the radiation of photons of energy $h \cdot \nu$ in thermal equilibrium (whatever the origin of this thermal equilibrium may be, for instance, the interaction of the photons with the heated walls of a black body).

---

[3] Planck, M. "Zur Theorie des Gesetzes der Energieverteilung im Normalspektrum," in *Verhandlungen der Deutschen Physikalischen Gesellschaft*, vol. 17, 1900, pp. 237–245.

One year earlier, Planck had already introduced this constant **h** and had obtained its correct order of magnitude.[4] Convinced of its fundamental importance, he proposed a completely new system of units where centimeter and second are replaced by length and time quantities defined uniquely by the constants **c** (the speed of light in vacuum), **G** (the constant of gravitational interactions),[5] and **h**, his new constant. He believed, rightfully as we shall find out, that these quantities have a universal significance, *for all periods and all civilizations, even extraterrestrial and not human, and might therefore be qualified as natural.*

He proposed the following quantities:

- *Planck length* or *Planck scale*: of the order of 10^{-33} centimeters[6]; a distance a trillion times smaller than the proton scale!
- *Planck time*: it is the travel time of light to span one Planck length; it is of the order of 10^{-43} seconds — a very, very short time indeed!

With Planck length and Planck time, we really are in the domain we call "the infinitely small." But what is the significance of such quasi-surrealistic quantities?

Planck did not interpret these quantities. How could he have interpreted them, since, at that time, quantum physics where h plays an essential role did not exist?! Neither did general relativity, which requires the constants c and G! And we did not know much about the dimensions of the subatomic world and practically nothing about cosmology. Planck displayed here an extraordinary scientific intuition, because these Planck quantities would get their significance in quantum physics and in general relativity.

Today, knowing that the constant h and the constants c and G characterize, respectively, quantum effects and general relativity, we may

---

[4] Planck, M. "Über irreversible Strahlungsvorgänge," in *Sitzungsberichte der Königlich Preußischen Akademie der Wissenschaften zu Berlin*, vol. 5, 1899, pp. 440–480 (in this paper, Planck refers to his constant as **b**).

[5] Remember Newton's law giving the force of attraction between two masses m and M separated by the distance R: $\mathbf{F = G \cdot (m \cdot M / R^2)}$.

[6] Planck length $= \mathbf{[G \cdot h / c^3]}^{1/2}$.

propose an interpretation of the Planck length and the Planck time: when classical general relativity, which studies the effects of physical objects on space and time, tests smaller and smaller dimensions, reaching space and time intervals of the order of the Planck scale, one may reasonably suppose that the quantum effects of gravity come into play. François adopts this viewpoint, which he considers unavoidable, but as we (as yet) do not have a reliable quantum theory of gravity, we may only suggest a blurring of the notions of space and time to make room for new notions that are still to be defined. An example of such a situation will be analyzed in Section VI.4.

Planck also defined a "temperature" based on the constants h, c, and G:

—   *the Planck temperature*: it is of the order of $10^{32}°K$, thousands of trillions times higher than the hottest stars![7]

One may calculate that photons emitted by a black body at the Planck temperature have a wavelength on the order of the Planck length. This result shows that the Planck temperature is also a limit, and in the absence of a quantum theory of gravitation, we cannot define the notion of temperature above Planck temperature.

This fact will have a very important consequence in cosmology, and particularly when the question of the origin of the universe arises.

To summarize, the notions of length and time may lose their operational significance below Planck length and Planck time, as does the notion of temperature above Planck temperature. All we can presently say is that the notions of space and time as currently used may objectively have only the status of concepts emerging from a different reality, or might become discrete, replacing the continuum that our old "common sense" once dictated …

Let us dream …

In any case, we see that the "infinitely small" and the "infinitely large" amply deserve their quotation marks, as the promise of an adventurous and deliciously thrilling research field.

---

[7]This temperature corresponds to the Planck energy $\approx 10^{19}$ GeV. Planck energy is related to the Planck (rest) mass by the relation $E=mc^2$.

VI.2. The Boson Goes Its Way … And François Goes His

1964–2012. A prophetic theory was elaborated in 1964, brilliantly confirmed in 2012. The public at large were soon imagining François Englert, Robert Brout, and Peter Higgs tied up in knots, on the lookout for nearly half a century.

Nothing could be further from the truth.

François and Robert never ceased to take an interest in the full range of the fundamental theoretical problems of modern physics.

Between 1964 and 1981, François carried on working in the area of "the infinitely small": quantum field theory, symmetry breaking, and strong interactions — alone or with Robert Brout, in collaboration or not with the researchers of the Pool de Physique, Marie-Françoise Thiry, Claude Truffin, Pierre Nicoletopoulos, Jean-Marie Frère, Harry Stern, and the French physicist Cirano De Dominicis, with whom François worked during the academic year 67–68 at Saclay in France.

Cirano De Dominicis, of the Commissariat à l'Energie atomique, at Saclay, five years older than François, was a major figure of the generation of young physicists who, after the Second World War, instigated the renewal of theoretical physics in France. He was a warm, easy-going fellow, displaying unbounded energy. His scientific achievements were impressive. His heart also leaned to the left, and he was active both politically and professionally within the scientific community. Quickly, François and Cirano became friends and together would experience memorable festive events.

This research work of François, mainly in what we call the "infinitely small," did not prevent him from making a successful incursion, in 1975 and 1976, into the domain of general relativity, in collaboration with Edgard Gunzig, Claude Truffin, and Paul Windey of the ULB and Raymond Gastmans of the KULeuven, where questions about the quantization of general relativity would be analyzed.

Then, two years later, François, Robert, and Edgard Gunzig, in collaboration with Philippe Spindel from the University of Mons — who had already been collaborating for a couple of years with the Groupe de Physique Théorique at the ULB — initiated fundamental research in

Cosmology that would result in a novel and enticing theory of the birth of the universe (Section VI.3).

This was also the time of the discovery, one year after the paper published by Robert Brout and François Englert, of the cosmological background radiation coming from all directions of space, filling the whole visible universe. Is it the remains of a singular explosive event that gave rise to the universe? This is posited as the Big Bang hypothesis. We shall come back to discuss this issue, but let us first point out that from this side of the "infinitely large," the BEH mechanism would open astonishing perspectives by proposing a scenario whereby the mass of massive elementary particles appeared "very shortly after" the onset of the universe, giving rise to its basic content as we know it.

VI.2.1. *Cosmology in the early seventies*

Despite the BEH mechanism that provides an explanation for the origin of elementary particle masses, and all the other great progress in the understanding of quantum phenomena, the birth of the universe itself remains an enigma. And if indeed there is a BEH mechanism to give the elementary particles their mass, when does it come on the scene? What do we really know about the first instants of the universe?

Concerning the BEH mechanism itself, despite its firm theoretical basis acquired in the early seventies, we should not forget that at that time, the BEH boson had not yet been detected. However, indirect experimental evidence, such as the discovery at CERN of the massive bosons of the electroweak theory (Section V.5), predicted by the BEH mechanism, kept the optimism alive.

The scientific revolution of the Renaissance brought us to discover the universality of physical laws, and that they govern all phenomena both on earth and in the heavens. Since then, we have come to know that the whole visible universe is composed of the same atoms as on earth, and that the Standard Model applies to the whole observable universe and its contents. In addition, it enables us to go very far back in time to explore its history. To understand this issue, one must appreciate two essential discoveries in astrophysics that occurred during the 20th century:

VI.2.2. *First great discovery: The universe is expanding*

In 1922, the Russian physicist Alexandre Friedmann showed that Einstein's general relativity involved the notion of an expanding universe, which meant that (on average) any two celestial objects move away from each other with a velocity proportional to the distance between them.

In 1927, the canon Georges Lemaître, lecturer at the Université Catholique de Louvain,[8] not only rediscovered this theory but also verified it by examining the distant stars and galaxies. He showed that, on average, these galaxies indeed recede at a speed proportional to their distances. Two years later, the American astronomer Edwin Hubble would confirm these results by improving the measurements' precision. The Friedmann–Lemaître expanding universe is today unanimously acknowledged. Later, in the early thirties, Lemaître then proposed what he called "the theory of the primeval atom" to describe the cosmological model of a universe expanding from a very dense and — still very mysterious — initial state. This model was once jokingly nicknamed "the Big Bang" by a physicist who believed in a stationary universe, and that tag has stuck.

However, although one might think that the expansion of the universe implies its birth at a moment determined by rewinding its history from its present stage to its origin, this could well prove incorrect. One may indeed imagine a continuous creation of matter that maintains the universe stationary despite its expansion, as this is also a solution compatible with the theory of general relativity.

François recalls that in 1960, when he was Robert Brout's research associate, he participated in a conference on Cosmology and was baffled by how heated the debate was between those arguing in favor of a universe emerging at a given moment and those defending a stationary universe, the latter accusing the former of attempting, under the guise of scientific considerations, to justify their theological convictions. But the issue would be resolved a few years later.

---

[8]The predecessor of the KU Leuven.

VI.2.3. *Second great discovery: Cosmological background radiation*

In 1965, two astronomers, Arno Penzias and Robert Wilson, while attempting with a Bell Telephone antenna to measure the intensity of the radio electromagnetic waves emitted by our galaxy outside the Milky Way, noticed a noise registered by the antenna no matter which way it was pointed. They first attributed it to a defect in their measuring device; they tried in vain to get rid of it. Informed by a preprint[9] from their "neighbor," Princeton University, discussing theoretical considerations favoring the existence of "ancient" cosmological radiation tracing back to the origin of the universe, they realized that they had just made a discovery of exceptional importance.

The analysis of this radiation showed that it originated from an ensemble of primordial photons in thermal equilibrium at a very high temperature, whose wavelengths span the Planck distribution curve characteristic of the temperature under consideration (Section III.2). This radiation followed the history of the universe since the appearance of primordial photons, cooling as the universe expanded. This history, experimentally verified, could not have happened in a stationary universe. This radiation that Georges Lemaître would poetically qualify as *the lost echo from the formation of the worlds,* the existence of which he somehow foreshadowed thirty years prior to its discovery, definitely confirmed the hypothesis of an expanding universe, and bolstered the defenders of the Big Bang.

VI.3. Cosmology and Big Bang

The discovery of the cosmological background radiation offered an answer to the timeless question: does the universe have an age, and if so, how old is it?

The measurements of the expansion velocities have enabled us to date the birth of the universe at about 13.8 billion years ago. According to

---

[9] Dicke, R., Peebles, J., Roll, P., and Wilkinson, D. "Cosmic Black-Body Radiation," in *Astrophysical Journal,* vol. 142, 1965, pp. 414–419.

classical physics, this must have happened at infinite temperature and at infinite density. This is the Big Bang singularity[10]: a dead end for science, as such a singularity cannot be analyzed in scientific terms. Let us for the moment forget this issue; we shall return to it later.

Taking into account the limit of the validity of the notion of temperature at Planck temperature (Section V.I) (around 10^{32}°K), let us confine our tracking back in time to that temperature[11]: this does not significantly affect the dating of subsequent events.

Starting from 10^{32}°K, the temperature decreases with expansion, and when it reaches 10^{16}°K, the universe is $10^{(-11)}$ seconds old. One is then within the range of energies reachable by present particle accelerators (the LHC at CERN), and we can describe by the experimental and theoretical data from the Standard Model the further evolution of the universe in detail, both in its content and in the evaluation of its expansion rate, the accelerators actually playing the role of time machines.

That is to say, in the vicinity of this temperature of 10^{16}°K, one encounters the electroweak phase transition where the elementary particles, bosons and fermions coupled to the BEH boson, previously massless, acquire their mass through the BEH mechanism. It is thus after $10^{(-11)}$ seconds that the BEH mechanism comes into play.

Let us continue our descent in temperature. After about 3 minutes, we encounter phenomena of capital importance, and among them, the formation of light nuclei, mainly helium nuclei. But we will have to wait nearly 400,000 years for the temperature to decrease down to about 3,000°K, to allow for the nuclei to bind with electrons, forming electrically neutral atoms. At that moment, the photons, which were in quasi-equilibrium because of their interactions with charged particles, cease to interact with matter, which is now neutral, and escape into the universe that has become transparent. They will travel unencumbered down to us, covering distances of billions of light years. During this journey, the spectrum of photons, namely, the wavelength content of light, will integrally conserve its Planck distribution, at the temperatures reached during their cooling

---

[10] i.e., the existence of a space-time region where certain quantities become infinite.

[11] Remember that Planck units are limits below or beyond which measurements become elusive due to quantum fluctuations.

due to expansion, down to the present time, where their temperature reaches 3°K. This homogeneity is amazing and will be explained in the next episode: no other measurement of the Planck curve has been obtained with comparable precision. Moreover, we shall see later that the very small fluctuations around that curve have an extraordinary significance: because of the tiny gravitational interactions, these fluctuations, whose origin we shall later discuss, will slowly evolve toward the great inhomogeneous structures of the universe, the galaxies, the stars with their planetary systems where life would develop.

VI.3.1. *But what can then be said about the Big Bang?*

François and Robert, as many others later, found the Big Bang theory less and less convincing even as it was practically unanimously adopted after the 1965 discovery of the cosmological background radiation, but that now appears to some as *a statement of perplexity and despair rather than the result of a logical argument* ...[12]

In 1977, they decided to thoroughly explore this question.

Let us try to "visualize" our universe and its evolution. The Planck quantities (here specified only by their order of magnitude) will help us to clarify the issues.

A brief recap:

— Planck length (scale) = 10^{-33} cm as the lower limit that may presently be unambiguously interpreted as a length.
— Planck time = 10^{-43} seconds as the lower limit of time span ...
— Planck temperature = 10^{32}°K. as the higher limit of temperatures ...
— A photon takes 10^{-43} seconds to travel a distance of 10^{-33} cm.

The radius of the present observable universe is of the order of 10^{28} cm, that is 10^{61} times the Planck length. This is the distance

---

[12] Brout, R., Englert, Fr., and Gunzig, E. "The Causal Universe," in *General Relativity and Gravitation*, vol. 10, 1979, pp. 1–6. This paper is a summary of their fundamental paper "The Creation of the Universe as a Quantum Phenomenon," in *Annals of Physics*, vol. 115, 1978, pp. 78–106.

that, from an observer within the universe, marks its cosmological horizon. The cosmological horizon limits the content of the universe to the objects that recede from us at a speed lower than or equal to the speed of light.

For all Planck distributions (Section III.2), the number of photons in a given volume is proportional to the cube of the temperature. One may thus calculate the number of background photons in the presently observable universe (where their temperature is 3°K). This yields, as an order of magnitude, 10^87 photons, a number that can be checked experimentally, because the number of the other photons emitted by stars and galaxies is negligible compared to the number of these background photons that populated the universe and satisfy with an impressive precision the Planck distribution. What volume did these photons occupy "in the beginning," when they were in temperature equilibrium at roughly the Planck temperature, with wavelengths comparable to the Planck scale (in accordance with our assumption about the significance of Planck units)? Taking this as (roughly) the original temperature, we get 10^(87) Planck volumes.[13] We should therefore imagine these 10^87 photons spreading over a distance of roughly 10^29 Planck lengths,[14] that is, over many billions of trillions times the Planck length! If we suppose that these photons were emitted by a Big Bang, their temperature equilibrium would then be very difficult to explain.

François and Robert did not think that the universe arose from a Big Bang, and for them, it was mandatory to search for a theory of the birth of the universe that described as closely as possible this challenging phenomenon.

It was precisely such a theory that was presented in 1978 by Robert Brout, François Englert, Edgard Gunzig, and Philip Spindel. As we will see, in the scenario they imagined, the problem of the classical initial conditions is avoided and the origin of the universe appears as a consequence of its quantum nature.

---

[13] The Planck volume is taken to be the Planck length cubed.

[14] $(10^{29})^3 = 10^{87}$.

VI.4. Genesis According to Brout, Englert, Gunzig, and Spindel (BEGS)

Robert Brout, François Englert, Edgard Gunzig, and Philippe Spindel would create the universe from a quantum fluctuation of the vacuum at the Planck scale.

VI.4.1. *What is vacuum?*

In classical physics, vacuum is emptiness. It contains nothing, no particle, no field, in particular no photon and no electromagnetic field. This is not the case in quantum physics, where the vacuum of quantum electrodynamics will necessarily contain, even in absence of photons, local fluctuations of the electromagnetic field imposed by the Heisenberg uncertainty relations (Section III.2). The quantum vacuum is not characterized by the total absence of particles and fields, but can be taken to be a state of zero energy. One may imagine the original space-time as filled by interacting fields fluctuating around zero, like an ocean disturbed by the quivering of particles and fields, referred to scientifically as the quantum fluctuations of the vacuum.

But then, where does the matter of our universe come from, such as we know it today, in particular the matter of the galaxies and of their stars? Where does its energy come from?

The answer put forward by Brout, Englert, and Gunzig was both "simple and elegant" and Spindel's contribution particularly informative.

VI.4.2. *Tracking the first quiver*

Brout, Englert, and Gunzig conceived the universe as a response to a local quantum vacuum fluctuation where matter was "seized"[15] by the gravitational field to which it was coupled. And this happened in the vicinity of the Planck scale, as would be shown by Spindel.

---

[15] "Seized" means that the fluctuation, instead of vanishing as fluctuations normally do, will trigger a physical process.

A general and extraordinary property of general relativity is essential here: in the cosmological context of a homogeneous expanding universe, the energy of the gravitational field is and remains exactly equal and opposite (negative) to the positive energy carried by the matter that generates it and that it generates in turn through its expansion.[16] The initial fluctuation will in this way create an expanding universe in full agreement with the fundamental law of energy conservation. As one may read in their 1978 paper[17]:

> How did mass appear ex nihilo, where does the energy come from? How can we answer this question if we do not appeal to theology? There exists a simple and elegant answer: when matter is created, it is carried by a gravitational field. The extraordinary feature is that the energy associated with this field is exactly equal and opposite to the energy carried by the matter which generates the field. Fundamental conservation laws are thus upheld.

The key to understanding this statement is hidden in the words "in the cosmological context." Indeed, the theorem of energy conservation that enables the creation of the universe from a zero-energy quantum vacuum, a creation quasi ex nihilo, is valid only if the created gravitational field forms an expanding homogeneous space. We shall later understand how, out of this original homogeneity, the huge inhomogeneities that are galaxies, their stars, the planetary system, and their inhabitants can arise.

VI.4.3. *The discovery of inflation*

The justification of the hypothesis proposed by Robert, François, and Edgard required considerable mathematical paraphernalia that we won't impose on the reader.

We shall however try to explain the origin of the matter created out of a vacuum fluctuation and how this hypothesis may render superfluous the Big Bang and its singularity.

---

[16] Matter creates gravitation, and gravitation creates matter; this is "retroaction," a phenomenon frequently encountered in physics.

[17] Brout, R., Englert, Fr., and Gunzig, E. "The Creation of the Universe as a Quantum Phenomenon," in *Annals of Physics*, vol. 115, 1978, pp. 78–106.

Let us say roughly (very roughly) that "before," there was no matter, no gravity, and zero energy. In this quantum vacuum, Brout, Englert, and Gunzig introduced the field of general relativity and a massive scalar field of mass m. A quantum fluctuation of this scalar field is then seized by the gravitational field to which it is coupled to form a macroscopic homogeneous medium that will develop according to the laws of classical general relativity interacting with a quantized scalar field. One finds that its expansion will be extremely rapid (exponentially rapid) and will create a dense medium of particles of mass m that will fill the homogeneous space of the nascent universe. During this creation phase, the total energy will stay equal to zero, because, as already mentioned, the negative energy carried by the gravitational field is compensated by the positive energy carried by the created matter.

This new world expands very coherently but this coherence will finally be lost as a result of matter disintegration and all hitherto neglected interactions (such as, for instance, the other components of the gravitational field). All these perturbations will destroy the initial coherence, provoking huge heating of the created matter that was initially cold. Creation then stops, and the universe further evolves from this primeval thermal state at a finite temperature.

But can we introduce the field of classical general relativity which cannot, like the scalar field to which it is coupled, be quantized and therefore can only be treated classically? Remember that, going back in time to scan the history of the universe, the effects of quantum gravity are expected to come into play only at very, very small distances (of the order of the Planck scale) while quantum effects in matter are already present when reaching the atomic scale. It is therefore legitimate to consider in cosmology the coupling of quantized matter to classical gravity.

The scalar field masses are *a priori* arbitrary, but Philippe Spindel succeeded in obtaining an exact formulation of matter creation[18] in the model proposed by Robert, François, and Edgard, and showed that the mass of the scalar field actually had to be taken in the vicinity of the

---

[18]Brout, R., Englert, Fr., and Spindel, P. "Cosmological Origin of the Grand-Unification Mass Scale," in *Physical Review Letters*, vol. 43, no. 6, 1979, pp. 417–420; Spindel, P. "Mass Formula in a Cosmogenesis Model," in *Physics Letters*, vol. 107B, no. 5, 1981, pp. 361–363.

Planck scale,[19] i.e., close to this space of initial "instants" governed by quantum gravity that we presently do not understand.

This result suggested that the birth of the universe also marked the emergence of classical gravity, and probably also of the notions of time and space. The scenario presented here indicated a possible path toward such emergence and its stabilization.

The present tale of "Genesis" obviates the Big Bang and its singularity. Starting from a quantum fluctuation of the vacuum, it unfolds in three stages:

— The emergence of classical gravity (general relativity) in the vicinity of the Planck scale.
— The creation of matter in a homogeneous space expanding exponentially fast that creates and stabilizes this classical gravity.
— The subsequent loss of coherence and its consequences, which induces an intense heating at very high temperature and marks the end of the exponential expansion.

This sets the initial conditions of the universe without requiring a Big Bang.

The mechanism that provoked a phase of exponential expansion would be rediscovered one year later in a different context by the Soviet Physicist Alexeï Starobinsky[20] and later, in the 1980s as a purely classical phenomenon, labeled inflation. We shall specify it as **primordial inflation,**[21] an inflation born in the vicinity of the Planck scale.

This primordial phase of the universe ends up at a very high temperature and the universe then evolves in a state of free expansion, at a

---

[19] A (rest)-mass, defined only, like the other Planck units, as a function of Planck constant h, gravitational constant G, and c, the velocity of light, with Planck energy $E = mc^2$.

[20] Starobinsky, A. "Spectrum of Relict Gravitational Radiation and the Early State of the Universe," in *Journal of Experimental and Theoretical Physics Letters*, vol. 30, 1979, pp. 682–685.

[21] Primordial inflation is a phenomenon of quantum creation of massive particles, which differs from classical inflation, developed in the 1980s, by the fact that here, the scale of the inflation is fixed in the vicinity of Planck scale, whereas for classical inflation, the scale can be arbitrarily chosen.

considerably slower rate, in agreement with the laws of general relativity —
a universe born without any huge explosion and coming of age under the
harmonious double guidance of the laws of quantum physics and general
relativity.

Primordial inflation will have considerable influence on the structure of the universe up to its present structure. It maintains the homogeneity required for the creation period, because any classical inhomogeneity is immediately washed out by the exponential expansion. But this is not the case for the quantum fluctuations that modulate exponential expansion: these are indeed unavoidable because, while being tiny, they are constantly recreated. Transmitted to the photons formed at the final stage of inflation, they will affect the radiation homogeneity some 400,000 years later. From there, the resulting inhomogeneities will during billions of years considerably amplify under the effect of gravitational attractions to give birth to the great inhomogeneous structures in agreement with all physical laws.[22] Galaxies, stars, and planets will form, evolve, ours in particular, leading finally to the emergence and the extraordinary development of biomass, namely, life.

Actually, the idea of a universe emerging from a quantum fluctuation was not new but the analysis of its consequences by Brout, Englert, Gunzig, and Spindel was new. The first suggestion of the quantum origin of the universe was made by Georges Lemaître in 1931 and was later proposed in more detail by Edward Tryon.[23] The theory of Brout, Englert, Gunzig, and Spindel of the quantum origin of the universe, as an attempt to eliminate the inconsistencies of the Big Bang, gracefully succeeded, starting with a primordial inflation and its quantum fluctuations, in accounting for the universe in all its complexities.

In 1978, the paper entitled "The Creation of the Universe as a Quantum Phenomenon" appeared in the review *Annals of Physics* and,

---

[22] Mukhanov, V.F. and Chibisov, G.V. "Quantum Fluctuations and the Non-singular Universe," in *Journal of Experimental and Theoretical Physics Letters*, vol. 33, 1981, pp. 532–535.

[23] Tryon, E. "Is the Universe a Vacuum Fluctuation?" in *Nature*, vol. 246, 1973, pp. 396–397.

a few months later, Brout, Englert, and Gunzig published a simplified summary.[24] For this essay about the physics of "the infinitely large," François, Robert, and Edgard would receive in 1978 the First Award of the "International Gravitational Contest" from the "Gravity Research Foundation" which honors the best essay on a subject related to Gravitation each year. The laureate paper then appears as the lead article in the annual volume of the *General Relativity and Gravitation Journal*.

Robert Brout considered this scenario of the creation of the universe from a quantum vacuum fluctuation causing a primordial inflation an important contribution to physics, even exceeding that of the BEH mechanism. In fact it was one of the subjects that he would go on to study in his future research.

And we cannot resist the pleasure of quoting this beautiful (French) anagram formulated by Etienne Klein and Jacques Perry-Salkow, reported by Jean-Pierre Luminet: *Le vide quantique est et reste la source réelle de l'espace-temps et de l'Univers"* which yields the anagram: *"Qu'est-ce que la Terre devant ces vallées supérieures de pétillement d'étoiles.*[25]

VI.4.4. *When Cosmology starts itching ...*

In the years that followed, François would keep exploring the origin of the universe and the primordial inflation hypothesis, alone and in collaboration with Robert Brout, Edgard Gunzig, Jean-Marie Frère, Claude Trufin, and Pascal Nardone, his ULB colleagues, Philipppe Spidel of the University of Mons, and Aharon Casher of the University of Tel Aviv.

---

[24] Brout, R., Englert, Fr., and Gunzig, E. "The Causal Universe," in *General Relativity and Gravitation*, vol. 10, 1979, pp. 1–6. A summary of their fundamental paper "The Creation of the Universe as a Quantum Phenomenon," published in *Annals of Physics*, vol. 115, 1978, pp. 78–106.

[25] Klein, É. and Perry-Salkow, J. *Anagrammes renversantes ou le Sens caché du monde*, Paris, Flammarion, 2011; Luminet, J.-P. *L'Écume de l'espace-temps*, Paris, Odile Jacob, 2020.

The quantum void is and will remain the genuine source of space-time and universe becomes *What is Earth when compared to these mighty valleys and sparkling stars.* Anyone to venture an equivalent anagram in English?

In a way, his research was a prelude to a project to which François would devote the greater part of his time starting in 1982: the attempt to reconcile quantum physics and general relativity, the grand question that continues, even to this day, to constrain the elaboration of a theory that would more completely describe the first instants of the universe's existence. Not to mention that surpassing what appears as a major conceptual difficulty might well revolutionize physics at every level.

VI.5. A Fertile Haven within the ULB

In 1980, Professor Jules Géhéniau retired. François stepped in to become director of the Mathematical Physics Department. Géhéniau chose this moment to convey his regrets at not having understood how innovative Brout and Englert had been, and at hesitating back then to set the department down that path. François agreed to succeed him provided that he not carry a heavy teaching load. Not a problem, his "Field Theory" class was gladly taken over by associate professor Christiane Schomblond, while he would continue giving classes at the Pool de Physique.

In the Mathematical Physics Department, which noted his arrival with some trepidation, they understood very quickly that François wasn't setting out to annoy anybody and that he would leave people free to pursue their work. He hadn't the slightest intention of gaining a "following"! The race to power did not interest him in the least. He detested that way of jockeying for position and had never applied for any kind of distinction, even though this is common practice among scientists, senior civil servants, military officials, ... and writers. He was after neither power nor money. His greatest satisfaction was the pleasure of doing research, preferably surrounded by fellow scientists, researchers "cut from the same cloth."

For many years, François directed master's theses and doctoral dissertations.

Paul Windey and Laurent Houart were two of the many students whose doctoral degrees were directed by François. The three researchers became great friends. François felt very close to these two young physicists who would share with him in many intellectual adventures, which will be addressed later. To cite but one, there was a feverishly inspired night in Paris during which the three of them produced an article.[26] Paul Windey would become a professor at Université Pierre-et-Marie-Curie (future Paris VI then Paris IV and finally Sorbonne Université). Laurent Houart would become a senior researcher at Belgium's National Fund for Scientific Research (FNRS), and at the ULB in 2004, he would be awarded the Prix De Donder of the Belgium Academy of Sciences.

François's circle of friends continued to widen. He received invitations to teach in the United States, Israel, Chili, and England, and took part in numerous colloquia. Master navigator of science, he was greeted by fellow travelers at each port of call. He would devote himself unstintingly to research from that point forward. His office at the ULB remained one of the focal points where the quest to understand the universe could be pursued in complete freedom and with a warm feeling of togetherness. And recognition by his peers would soon follow.

VI.5.1. *The Prix Francqui*

The Prix Francqui is the most important scientific distinction in Belgium, awarded annually by the Francqui Foundation[27] in the fields of Hard Sciences, the Humanities, and the Biological and Medical Sciences. Candidates can be vetted either by two members of the Belgian Royal Academy or by a former prize winner. The Foundation Board awards the prize upon a motion by an international jury.

In May 1982, François was awarded the Prix Francqui by a jury headed that year by Professor Abdus Salam (Section VII.2):

> *In consideration of his contribution to a theoretical understanding of the phenomenon of broken symmetry in fundamental physics where, along*

---

[26] Englert, F., Houart, L., and Windey, P. "Black Hole Entropy Can Be Smaller than A/4," in *Physics Letters,* vol. B372, 1996.

[27] It was Émile Francqui and Herbert Hoover who proposed the creation of the Foundation whose purpose would be to further "the development of higher education and scientific research in Belgium." The Francqui Foundation was created by the Royal Academy on 25 February 1932 as an institution of public utility.

with R. Brout, he was the first to show that spontaneous broken symmetries produced masses for gauge particles,

In consideration as well of the extension of his contributions into other fields such as solid state physics, statistical mechanics, quantum field theory, general relativity and cosmology,

And finally, in consideration of the originality and fundamental importance of his work, the Jury has decided to award the 1982 Prix Francqui to Mr. François Englert, Professor at the l'Université Libre de Bruxelles.

The prescient initial reaction by Papa Englert was, "Fine, and now hurry up for the Nobel Prize, I'm not getting any younger!"

VI.5.2. *A speech sets off storm*

As is customary, the recipient of the prize gives a speech responding to that of the President of the Francqui Foundation. François began his address according to expected standards:

It is with joy and pride that I express my appreciation for the honor you bestow by awarding me the Prix Francqui.

The work that you are thus rewarding is the result of a collaboration of over twenty years between Professor Robert Brout, an American, and myself.

[...]

We wish to express our special gratitude to Professor Paul Glansdorff, at the Université de Bruxelles, who welcomed us so warmly into his department that was such a faithful reflection of his humanism and knowledge.

[...]

Your Majesty,
Mr. President,
Ladies and Gentlemen,

What followed was a brief look at the research François carried out and a presentation, both scientific and philosophical, of basic questions that all physicists must confront.

Then came a somewhat longwinded discursion.

And there, the tone shifted dramatically.

François could no longer suppress his concern and exasperation at the way the academic world in Belgium operated. Before an audience of dignitaries, deciders, renowned scientists, in the presence of the ULB Rector and the King himself, he felt it his duty to sound the alarm:

All of our work fits into the overall prospective of developing knowledge whose field of application is ever expanding. Yet, I feel a certain uneasiness as I come to close my address here, because I fear it will contribute to fostering the illusion that my country might for a while still take part in this development and benefit from the findings that result. For this reason, I cannot keep a complicit silence as I watch my country move toward the point of no return into a long-lasting state of scientific underdevelopment. The fact that nearly all our top researchers have left the country is significantly reducing our fundamental research potential in inverse proportion to the mediocrity that has overtaken our scientific institutions. This cultural decline should not be seen as the direct result of reductions in available funds, but rather of incompetence and a lack of courage on the part of decision-makers who have failed to make academic appointments based on the sole criterion that matters at the university level: quality.

Consternation among the academic ranks.

Praise from King Baudouin who had listened to the speech with rapt attention: "A very beautiful speech."

He then called his adjutant,

Go get me the ULB Rector.

And when the Rector arrived, the King attacked point blank.

So, what does this all mean?

The rector stammered the conventional excuse:

Your Majesty, we have inherited a difficult situation ... and ...

François was uneasy, having sensed a heavy disapproval among the public composed mainly of bigwigs from his alma mater. And for a long time hence, he would go on feeling, like a chilly wind to his back, the malicious gaze of certain colleagues. Between François and the ULB, there would prevail a deep antagonism until the announcement of the Nobel Prize.

But François was not the only invaluable scientist that the ULB had decided to snub. A number of internationally reputed colleagues simply disappeared one day from the ULB, to the astonishment of their students, because the institution was plainly doing nothing to make them want to stay. Most would go on to brilliant careers abroad at the world's most prestigious universities.

Still, a Prix Francqui is worth celebrating.

Danielle Vindal, whom François married in 1971, decided to throw a huge party in his honor at the Uccle house on Rue des Pêcheurs, a house that would soon swallow up half of the prize money to make it habitable. For soon after they purchased it, the new owners would make the horrifying discovery that the house was not connected to the sewage system when cesspool started to overflow into the street, to the great displeasure of the neighbors, including an absolutely furious restaurant owner who, it would turn out, had something to do with the crisis!

Episode VII

Coming Together and Breaking Apart

Mon beau navire ô ma mémoire
Avons-nous assez navigué
Dans une onde mauvaise à boire
Avons-nous assez divagué
De la belle aube au triste soir.[1]

Apollinaire (*La Chanson du Mal-Aimé*)

VII.1. Ghosts of the Past …

As we have seen, François's life features an uncanny superposing of two discordant trajectories: one, the path of a scientific researcher, striding down a beautiful tree-lined avenue toward a chosen objective, and then the other, François's interior life, a steep trail strewn with stones and thorny underbrush, *full of sound and fury*. The hidden child is still there, deep within him, with all the anxiety, all the nightmares. There is something else gnawing at him, a slowly widening gap, a dizzying rift: the total erasure of his family's entire past, an erasure orchestrated by his

---

[1] My fine ship O my memory/Have we not sailed enough/In waters unfit to drink/Have we not strayed far enough/From beautiful dawn to unhappy eve.

159

own parents, clearly despite themselves, already dating back to before the war, for they never told François anything about their life in Poland. And after the war, forgetting became the most basic condition of survival. Never did the parents desire or find the words to evoke the fate of those left behind. Nothing from the family memory was passed down. If this erasure did "save" the parents, it did not save François, forever dispossessed of his roots, deprived of any possibility of going back in time. François was a child of an obscure land shrouded in fog, without geography or tutelary figures. To be "rooted" is a reassurance, an affirmation that we can legitimately claim who we are, where we are. For François, whenever something particularly painful arises, he literally feels the ground shift beneath his feet and falls back into the uneasy limbo of his obliterated past. A stranger in the world and a stranger to himself.

VII.1.1. *The end of a collaboration*

After their work on the origin of the universe and in particular on their hypothesis of a quasi-exponential primal expansion (later named primordial inflation), François and Robert further honed that hypothesis along with Philippe Spindel, and at the Pool de Physique, and obtained promising results.

But after that, François and Robert's paths would slowly begin to diverge.

François choose to devote himself chiefly to the development of more speculative theories that could lead to a quantum theory of gravitation, such as string theory[2] and supergravity, whose classical limits are generalizations of Einstein's general relativity.

François and Robert Brout had invited a Dutch physicist, Bernard de Wit, to the Pool de Physique for a class on supergravity. De Wit,

---

[2] String theory involves an attempt at unifying quantum physics and general relativity. Here, elementary particles are replaced by objects of a dimension called "strings." This implies spaces in a large number of dimensions. As with other theories, such as superstring theory and supergravity theory, these extra dimensions are introduced for reasons of mathematical consistency.

a researcher known for his infectious enthusiasm, prepared a one-hour lecture, but would extend his presentation over the next two days before an audience hanging onto his every word, interrupted only by the back-and-forth of lively discussions. This event would mark the beginning of a fruitful collaboration and a long friendship between Bernard de Wit and François, for whom quantum gravitation was the only hope for a rigorous approach to how the universe emerged, and where it is going. (We shall be returning to this huge physics question in Section VII.5.)

While François was following this path he had set for himself, Robert Brout was pursuing his study of elementary particles. He would next focus on the meaning of time in cosmology and write an important paper on black holes[3] within the framework of field theory. Meanwhile, François would also be studying black holes but in the more speculative context of string theory. Like Robert, he was working with people at the Pool de Physique, but also, as we shall see, with a more far-flung array of collaborators.

Despite this fork in the road, their friendship remained intact, but this break, however gradual, did destabilize François somewhat. He missed the warmly intense experience of shared work that he had so enjoyed with Robert. And right when he was facing the loss of this precious connection, he had to confront a much more painful break-up.

VII.1.2. *A relationship ends*

By 1980, François and Esther had already been separated for over ten years, a mutually agreed upon decision, and they had stayed friends. François was granted custody of the two daughters, while Esther kept Georges. François had been living on Rue General Patton with his partner Danielle, whom he would marry in 1971. Danielle and François would have two daughters together, Sarah, born in 1973, and Hélène in 1978. The friendship among François's five children, Michèle, Anne, Georges, Sarah and Hélène, was a very cordial one. François saw them on a regular basis, whether together or separately.

---

[3] In astrophysics, a black hole is a celestial object that has a gravitational field so strong that neither light nor any object can escape it.

But during the 1980s, François and Danielle's marriage was growing brittle and finally reached the breaking point. They separated in 1983 and would divorce in 1985.

François took this break-up very hard. Not only was he losing his partner but also the support of a warm and reassuring setting where everyone in the entire extended family had their place. Suddenly reemerged the specter of loneliness, abandonment, and anxiety experienced as a hidden child, "forgotten" by his near and dear, where he was once again engulfed by that feeling of being forever absent from his own story. He watched as his life was turned upside down.

He knew that only by throwing himself into his work and cultivating new friendships among his fellow scientists could he overcome his distress.

He was alone in Brussels, however, when in 1982 he started his new research project by first of all shifting toward supergravity, the theory that he and Robert Brout had been "initiated" to by Dutch physicist Bernard de Wit during his landmark visit to the ULB.

François had discovered in the library of Professor Paul Glansdorff the complete works of the great French mathematician Elie Cartan. Inspired by Cartan's study of parallelisms in curved spaces[4] and by his demonstration that a 7-dimensional sphere allows two exceptional parallelisms, François, with his peculiar intuition that short-circuited deductive thought, posited that these two parallelisms discovered by Cartan in the 1920s provided solutions to the very contemporary 11-dimensional supergravity!

The paper was published as a pre-print by CERN before appearing in *Physics Letters*[5] — during this period, François had several short stints at the CERN Department of Theoretical Physics. The paper would be reinterpreted in terms of octonions (a generalization of the notion of number) by Marianne Rooman in her dissertation directed by François at the ULB, and in the United States by Feza Gürsey and Chia-Hsiung Tze.

---

[4] Elie Cartan (1869–1951), in "La géométrie des espaces de Riemann" (1925). To present his study on parallelisms is beyond the scope of this book.

[5] Englert, Fr. "Spontaneous Compactification of 11-dimensional Supergravity," in *Physics Letters*, vol. 119B, 1982, pp. 339–342.

At CERN, François met Hermann Nicolai, an eminent physicist who also proved to be an excellent pianist, the future director of the Max Planck Institutes, and met up again with Bernard de Wit. Hermann, Bernard, and François together analyzed the problems of supergravity and pursued, along with other visitors to CERN, François's work on the 7-sphere. The findings obtained would make François famous throughout the world in this new area of research.

When François was invited to Princeton and to the International Center for Theoretical Physics (ICTP) in Trieste to talk about his work, he would be greeted with a warm *Welcome to the club*! by Daniel Freedman, professor at MIT and one of the architects of supergravity, and François was, of course, enchanted!

VII.1.3. *CERN*

When one hears the acronym CERN, one thinks immediately of the huge laboratory where hundreds of experimental physicists and highly specialized technicians are busy at work. Indeed, the European Organization for Nuclear Research is one of the largest and most prestigious laboratories in the world. Their tools include accelerators and particle detectors, and the main goal is to probe the basic make-up of matter. In other words, CERN's vocation is fundamental physics, i.e., to discover the laws of nature, to better understand what the universe consists of and how it functions. The presence within the institution of theoretical physicists is therefore essential. Free-flowing communication between theorists and experimenters is key to the success of the CERN programs. It provides for the constant testing of theory, just as it provides theory with "fresh" experimental data in order to explore new channels of thought.

As soon as CERN was created in 1954, a Division of Theoretical Studies was envisioned, one of first directors of which was to be the Belgian theoretical physicist Léon van Hove, professor at the ULB (Prix Francqui 1958) and CERN director from 1976 to 1980.

This section, today simply called the Theory Department, would develop in collaboration with CERN experimenters and would become a major center for groundbreaking theoretical research. It includes a restricted number of tenured theorists but also welcomes visiting physicists or under temporary contract from among the most eminent in the world to come work alongside the permanent staff. International colloquia are regularly organized on the most cutting-edge subjects dealing with both the "infinitely small" (particle physics) and the "infinitely large" (cosmology).

Starting in 1982, François was one of the regular visitors. An intense collaboration would develop between François and the Theory Department. A first publication emerged from this joint effort in 1982, followed the same year by a second (pre)publication with his ULB student Bernard Biran, Bernard de Wit, and Hermann Nicolai. Hermann and François often got together, to their mutual delight, at the ULB and the Gölm Max Planck Institute (Potsdam).

Then followed, to great fanfare, several publications on 11-dimensional supergravity and the 7-sphere, with such mind-bending titles as "The Fluctuating Seven-Sphere in Eleven-Dimensional Supergravity" and "Supersymmetry Breaking by Torsion and the Ricci-flat Squashed Seven-Spheres." Four papers in 1983 alone! They dealt essentially with speculative physics, off the beaten track, where intuition and imagination played a major role, and which were based on extremely sophisticated mathematics. They were all joint publications from the new group that had formed around François, including Bernard Biran, Bernard de Wit, Hermann Nicolai, Philippe Spindel, Aharon Casher from Tel-Aviv, Anne Taormina, and Marianne Rooman. These last two women would later become great friends of his, a subject to be addressed in Episode X.

In 1985, François was invited to work in the CERN Department of Theoretical Physics. This meant he would finally be earning a decent salary, but given his family situation — two exes and five children — he preferred to maintain residence in Brussels and continue teaching at the ULB while working at CERN. Thus, his routine would consist of spending his weekends in Brussels, teaching at the ULB on Monday, and then heading to Geneva.

CERN fulfilled his dream of intensifying the international dimension of his work, something he had always hoped to accomplish, especially given that the ULB's chilly attitude toward him had not warmed over the years.

Meanwhile,

Seeing Pakistanis collaborating with Israelis, Iranians with Americans, seeing fifty nationalities coexisting in harmony, far from the world in turmoil, is very heartening. CERN should be awarded the Nobel Peace Prize. Beyond the extraordinary technological developments it has achieved, it's the image of this rallying of humanity that is so amazing.[6]

Circles of friends would gradually form around François, who would find inspiration and comfort in both scientific and affective exchanges.

One through-line in François's life has been certainly his fondness for work-related friendships. As he himself avers, *I find interchange and partnership with other researchers to be essential. The ideas flow, it's the most fun part of discovery. For me, friendship in research is a huge motivating factor.*

How can one not think of the lovely Martin Klein text[7] on the friendship between Albert Einstein, Niels Bohr, and Paul Ehrenfest in the years following the famous 1927 Solvay Conference (Section III.2):

The warm understanding and strong bonds of sympathy established then between Bohr and Einstein sustained them through another thirty years of occasional meetings and deepening differences in their views on physics. Without this solid base of friendship, the Bohr-Einstein dialogue might not have continued to be a model of civilized disagreement over fundamental scientific issues. Without Ehrenfest's catalytic role, that friendship of Bohr and Einstein might never have achieved its full strength. Each of the three men felt revived, cheered and refreshed after contact with any of the others. Each felt, as Einstein had once written to Bohr, that the mere presence of the others was a source of joy.[8]

L'école buissonnière: **Wanderings**

With his first wife Esther, François had lived on Avenue de l'Armée. With his new partner Danielle, it was first on Rue General Patton, then Rue

---

[6]Interview with Guy Duplat, *La Libre* 4 July 2012.

[7]Klein, M.J. American mathematician, physicist, and historian of science (1923–2009).

[8]*The Lesson of Quantum Theory*, J. de Boer, E. Dal and O. Ulfbeck (eds.), © Elsevier Science Publishers B.V., 1986, p. 339.

Colonel Chaltin that they lived together, in an apartment previously occupied by his friend and colleague Harry Stern.

Avenue de l'Armée

Rue General Patton

Rue Colonel Chaltin

There would seem to be a military theme …

In 2015, François and his third wife Mira would find themselves on a street whose name commemorated — finally — a man of science, the astronomer François Folie, and such an overt nod to François's motto (Section IX.2)!

After breaking up with Danielle, François went through a period of great confusion and wandering.

One morning in the summer of 1983, François left home, carrying a small suitcase containing some clothes and other vital necessities, and sheets of paper filled with equations. He was hoping to find refuge at his brother Marc's, who indeed had agreed to let him stay for a while. But as he was leaving, right at his doorstep, he came face to face with Annette Deletaille, who was just stopping by to say hello. François explained the situation to her, and Annette took him back to her place, cheered him up and encouraged him to put his misfortunes into perspective. Annette, who had always been an especially resilient woman, made a point of heartening people by imparting her own moral strength:

Look at the stars, François, our problems are so tiny compared to the stars.

During a parent–teacher meeting at the Decroly School, François met Patricia (nicknamed Titine), the daughter of Safford Cape, an American composer and musicologist, native of Denver, living since 1925 in Brussels where he founded the Ensemble Pro Musica Antiqua. He had a smart and charming wife and two whimsical daughters, Titine and Miquette, who married the brother of Jean Orloff; he also had a son, the artist friend of Georges Miedzianagora, and who would fall in love with Noémie, the daughter of philosophy professor Chaim Perelman.

Titine adored animals, and lived surrounded by Muscovy ducks that terrorized her dog, and later, by geese and a goat. One day, when her alcoholic husband had threatened her with a gun, she had fled their home accompanied by her beloved geese.

François was desperately searching for lodgings where his children could also stay with him, but he was short on funds for the moment. After a few weeks spent in a kind of hovel lent to him by Françoise Henrion,[9] furnished only with a wood-burning stove and, for a mattress, a cat bed that François salvaged at Danielle's, he finally found an apartment to rent on Rue de Vergnies that had a room for Sarah and Hélène. His friend Nathan Clumeck helped him move in, and brought him several space heaters. Jean-Marie Bertrand, the husband of Jocelyne, a teacher at Decroly, brought him a mattress, a bit worse for wear, but he had to make do … François would live there alone, but he could rely on Titine who would always welcome him with open arms in the superbly spacious house that she inherited from her father and which she was sharing with her mother, at Clos du Vert Chasseur in Uccle. Likewise, during his stay in Geneva in 1985–86, it was again at Titine's, surrounded by geese, goats and ducks, that he would spend his Brussels weekends.

Jean-Marie Bertrand

Jean-Marie Bertrand, the mattress provider, was a fellow who very early joined the merchant marine. He loved to tell how the captain treated his crew: "Here's how things are here: I'm the boss, then there are the officers, then everybody else, then the dog, then there's you."

François found this flamboyant colossus very likeable.

They would take a number of getaway trips together, to Italy, Spain, Portugal, and later, to Israel.

Jocelyne and Jean-Marie had been separated for some time, and François was no longer with Danielle. The two divorcees planned their school holidays so that the children could all come on vacation together.

For the trip to Italy, the two friends planned to meet at the Novotel in Bologna, by the swimming pool, at precisely 8:30 pm. And Jean-Marie was there, at exactly the appointed hour with his children Gilles and

---

[9]Françoise Henrion, wife of Joseph Henrion, one of the creators of the art brut museum, *Arts et Marges*.

Violette, and François with Sarah and Hélène. Ignoring complaints from hotel management, Jean-Marie gave Sarah a diving lesson that same night, an event that she would never forget.

They next headed south and stopped at Lake Trasimeno, near Perugia. There, they found a huge fishing contest being held.

Jean-Marie was in his element. The day before the opening, the little gang went down to the lake. In less than an hour, Jean-Marie had caught a dozen fish with the children's help and the most rudimentary gear: a simple stick, some string, a little hook, and some worms collected by the children. The next day was the contest. Not a single contestant caught any fish! But Jean-Marie had discreetly given all his fish to one of the fisherman, who was applauded for this miraculous catch. But it did not escape the notice of the jury that these fish did not look especially fresh, and the hoax was exposed. A huge scandal erupted, which the two accomplices found hilarious… before slipping away with their children.

A few years later, there was a long trip through Spain and Portugal, with Jean-Marie, Gilles, and Violette, and François with four children in tow: Sarah and Hélène, Itamar, the youngest son of Joseph Katz, and Antoinette, one of Paul Glansdorff's grandchildren. Their trek took them through Evora and Lisbon, which the little tribe explored in detail, including the extraordinary Santa Justa elevator,[10] and ended in southern Portugal where François had rented a house. There were wacky moments throughout, like when Jean-Marie, having transformed the car into a Charles Trenet festival, turned the steering wheel over to his 14-year-old son Gilles.

Jean-Marie loved moving into new houses. He bought rundown properties and renovated them with lots of style. François recalls a white house whose paint was mixed with a methylene blue base, so that the house magically turned blue when it was rainy. Amazing! And Jean-Marie took advantage of these house moves to implement his unusual concepts of education. Before undertaking his nth renovation, he urged his children to demolish anything still left standing in the house. "Come on, smash everything and throw it out the windows!" No sooner said than done, amidst cries of delight!

---

[10] Which was designed by Raoul Mesnier du Ponsard and not Gustave Eiffel, as many tourist guidebooks claim.

Further down the road, this tireless and multifaceted whirlwind of a man would earn a degree from the ULB in urbanism and land surveying, and thanks to his bank loans, he was able to purchase apartments to renovate and flip for a profit. With his son Gilles, he went on to open a real estate office, and in the process, acquired an art gallery in Brussels.

The trials and tribulations of a physicist in the land of Uccle

Life is not a bed of roses

By 1987, François was back in Brussels after his tenure at CERN. He moved by himself into an apartment in Uccle, Rue Groeselenberg, one floor up. He rented until 1988, when he purchased the property.

The people living in this three-story building were involved in a series of episodes with François, not always pleasant but never unsurprising. There was no denying that throughout his entire life, François often got mixed up in some rather peculiar happenings. Was it mere coincidence, or a character feature that made him more receptive to the quirkier side, willingly seeking out situations that he anticipated, consciously or otherwise, would lead to interesting encounters? The non-conformist in him naturally cultivates a curiosity for people on the margins, and we know that he experiences great pleasure in sharing their adventures, with all the twists and turns they imply. But certain chapters in his life still remain shrouded in mystery.

The early years on Rue Groeselenberg passed uneventfully. After their divorce in 1985, Danielle got custody of their daughters, Sarah and Hélène, who went to stay with their father on a regular basis. All was well. Then came the moment when Sarah, the eldest, was old enough to have her first party with her friends, *chez* François. A little noisy gathering, so much so that a number of neighbors complained to François about the undue disturbance… in the middle of the work week, no less!

When François asked the partyers to lower the volume on the music, the teens raged on, pretending not to hear him. Exasperated by this music that was bursting his eardrums, he fled the apartment. He suddenly got the notion to seek refuge with the "lady upstairs" whom he had seen but never properly met. All he knew was that she had two children.

The door opened, and a pretty woman of a certain age stood on the threshold.

François tried to explain: his teenage daughter's first party, the infernal racket …

Very welcoming, the lady asked him in, and suggested they have a glass of whisky out on the balcony.

Conversation ensued.

The lady, a bit tipsy, stated bluntly that she was a *Madame.*

François was somewhat taken aback, but what's wrong with that, after all?

From the balcony, François glimpsed a well-kept flat, Madame's professional activities clearly took place elsewhere.

They continued to chat, naturally addressing the subject of their children. Her expression turning serious, the lady explained how important it was for her that her children did well at school.

As François was about to leave, she kindly declared:

"My dear sir, you will always be welcome here!"

And indeed he was. The lady felt sorry for him, a bachelor, and would make him a meal from time to time. A niece of hers who came to stay with her aunt told him that she was a prostitute. She was very nice indeed, and became friends with François. She even brought him back a gift from a trip to Israel, cassettes of songs in Hebrew.

But events took a turn. The niece's pimp, a decent man without a respectable profession, believed his livelihood was immoral and was hoping to turn his life around, but to make such a change required money, and to get money, there was only one way, he said: holding up a bank! Irrefutable logic.

And the guy did actually hold up a bank!

Yes, he robbed a bank, but unfortunately for him, he blew his chance and got caught and ended up in prison. Awaiting further news, the niece burst into tears, utterly hopeless.

But a few days later, he showed up at the apartment: he had somehow managed to escape!

The victory was short-lived, however, for the police were investigating in the building. The lovebirds had to hide. The police didn't find them this time.

A few days later, the two fugitives came knocking at François's door. The niece spoke first. "So here's the story, we've decided to turn our lives around, no more delinquency, we're going to start over; he's going to surrender to the police, and I'm going to give up prostitution. And we've decided to get married!" François, delighted at all these fine resolutions, offered to be a witness at their wedding. The repentant couple perked up again, a heartwarming sight.

But when the former pimp on the lam turned himself in to the police, they couldn't find his file, which had somehow been mislaid. They couldn't detain him in those circumstances, and were compelled to release him!

Such an improbable twist called for a toast! They had a party at François's place.

The next day, there was a deluge of phone calls from the landlord to the tenants following complaints by a number of neighbors about all the commotion caused by the police investigation, and then about the alcohol-soaked "ceremony" in honor of the lost sheep returning to the fold. There were also complaints following rumors starting to circulate about a strange "professor" who paid visits to the lady upstairs, and who danced with her ill-reputed niece at little impromptu parties. A bearded fellow in sweaters worn at the elbows, he could only be a shady character …

But things did eventually calm down.

And François continued to pay visits to *Madame*, the lady upstairs, and to attend her parties where, among swirling hashish smoke and whiskey vapors, friends mingled, most of whom were from a certain Brussels underworld.

They lovingly adopted François into their circle … which appealed to the rebel within him, to rub shoulders with these rather debonair dilettantes in crime. But when he spoke up after one of them made some racist remark, he got this utterly illogical response: "Racism is like the Arabs, it shouldn't exist." François was left speechless.

Later, François would see the niece and her pimp when they invited him to their cozy little house. He would always remember them warmly for their kindness to him.

As it turned out, the niece was the daughter of a good Uccle family that she had escaped from to "live her life." Her brother was a veterinarian, a friend of Jean-Marie Bertrand's, whom we have already met — it's a small world.

And every so often, the world does offer opportunities to men and women of good will: *Madame*'s children went on to study at university.

By 1988, building management problems were mounting. Simple technical issues took on epic proportions because of the absolutely insufferable character of the third-floor, garden-facing tenant. Meetings of co-owners devolved into psychodramas.

In 2003, the second-floor, street-viewing apartment was occupied by a woman placed under court supervision and her two daughters, both budding delinquents. The mother claimed to be Muslim, but would switch in a flash from niqab to blond wig and plunging neckline, while her daughters made a racket roller-skating around the apartment. One of her lovers was a well-built African man, perfectly charming but a bit too casual, while another was a brawler. The building manager, also a bit racist, enjoyed citing an adage of his own confection: *Arabs are violent, Africans are casual.* Incessant squabbles caused by the husband attempting to win back his wife put the neighbors in a terrible state. One evening, François discovered his garage filled to the ceiling with an array of assorted objects. The African lover soon showed up to apologize. "Oh, I thought the garage was for everybody to use…" As for her electric bills, the tenant had no intention of paying them. When François refused to let the little family hook up to his power meter, despite the delicious honey cakes the mother regularly brought him, she decided, without asking this time, to hook up to the meter belonging to a Creole woman living up on the third floor, street side. But the pirated connection was hardly discreet, since you could see from the street the electric wires snaking from one balcony to the other, to the total indifference of the young Creole who, elegantly dressed, walked her dog every morning, and every evening could be seen coming home in a taxi dead drunk. On occasion, François had to carry her up to her apartment or call for an ambulance.

Fights among some of these strange tenants were becoming more frequent. One day, blood was found on the stairs. Was it human blood, or that of a recently missing cat, killed in retaliation for some other dispute? Worse, an increasingly nauseating odor had been emanating from the stairwell. Garbage must certainly have been dumped somewhere.

François, by now, had run out of patience and went before the Justice of the Peace to file a complaint against the tenant on the second floor. The judge, a woman known for her left-leaning politics, approached the case with wisdom and intelligence, and offered to mediate. At the hearing, she took advantage of the absence of the defense lawyers to replace them with lawyers from her circle who quickly found another apartment for the defendant. The eviction could now move ahead.

When the landlord regained possession of his apartment, he found an absolute disaster zone: all the doors were gone, probably burned for heat, and the ceiling had collapsed after the "casual" installation of a chandelier that proved too heavy.

As for the young alcoholic Creole, it had become clear that the foul odors had been emanating from her flat. François wrote to the police to alert them of a "sanitation emergency." The reaction was swift. In under 24 hours, the young woman was gone and the landlord, ordered to clean up the mess, discovered his flat in an indescribable state of filth: rotting garbage bags, dog feces all over the floor, and the fridge full of greenish concretions too far gone to identify.

VII.2. An International Network

During and after his time at CERN, François was surrounded by a highly eclectic circle of physicists, some of whom were already well known: Hermann Nicolai, Bernard de Wit, Philippe Spindel, three of François's former Ph.D. students, Bernard Biran, Paul Windey, and Marianne Rooman, Jean-Marie Frère, Aharon Casher and Yakir Aharonov from the University of Tel-Aviv, Anne Taormina, now professor at the University of Durham and her Thai student Auttakit Chattaraputi, André Neveu from the Ecole Normale Supérieure in Paris and later at CERN, Tsvi Piran from the University of Jerusalem, Jean Orloff from Clermont-Ferrand, former student of François and Robert, Paola Zizzi from the University of Padua, Peter West from Kings College London, and many others to be mentioned further on, and who were regularly and warmly welcomed into the Pool de Physique.

This circle corresponded with other universities throughout the world, and in Belgium as well, notably with the KU Leuven. Their collaboration would produce a great number of papers in leading-edge fields such as string theory, supergravity, fractals,[11] and infinite dimensional Lie alge-bras[12] called Kac-Moody. This was the realm of physics that François had already approached before his time at CERN and that attempted to reconcile general relativity and quantum physics.

At CERN, where he frequently returned, every visit involved meeting someone new, such as Professor Jnanadeva Maharana from the Institute of Physics in Bhubaneswar, India, who left François and Mira with an unforgettable gastronomic memory during one of his visits to Meyrin, the little town that housed most of CERN and its residents where a majority of the center's collaborators and visitors lodged during their stay. One evening, as François and Mira were leaving their apartment, they were suddenly enveloped by the fragrance of something deliciously exotic. François immediately identified the source of this culinary wonder: Professor Maharana's apartment, right across from theirs. He couldn't help himself, and knocked at the door, ostensibly to ask him for the address of a good local restaurant. The professor wasn't fooled for a minute but, delighted for the impromptu occasion, Maharana and his wife, making no fuss, invited François and Mira to share their tasty meal.

In addition, François took part enthusiastically in numerous summer institutes in theoretical physics that were cropping up all over the place, and often in magical venues: the island of Corfu, Erice in Sicily, Cargèse in Corsica, Les Houches at the foot of Mont Blanc, or places in Florida, California, South Carolina, India, Vietnam, and Japan. Sometimes, he went by himself, other times with Robert Brout, both accompanied by their spouses. He was also a regular at the International Center for Theoretical Physics in Trieste, founded by the Pakistani physicist Abdus Salam, co-recipient, along with Sheldon Glashow and Steven Weinberg, of the Nobel Prize in Physics in 1979. This center, which aspired to be a model of cooperation among scientists in the developing world and in more scientifically advanced countries, benefited from the support of the

---

[11] A fractal is a geometric object defined particularly by this property: the whole is the same as each of its parts. An intuitive example: a fern leaf.

[12] The notions of Lie groups and Lie algebras (named after the mathematician Sophus Lie) are used abundantly in quantum physics. Explaining their mathematical bases does not fall within the scope of this book.

Italian government, the United Nations, and UNESCO. Its activities involved classes, conferences, and summer institutes. François has fond memories of the time he met Abdus Salam, the wunderkind of his native Pakistan, Cambridge graduate, respectful of Muslim tradition: he had two wives, one a cousin, the other a British biochemist and crystallographer, professor at Oxford. Both attended the Nobel ceremony, but the somewhat bewildered Swedish officials made sure they were sitting at opposite ends of the auditorium. As for the monetary prize itself, Abdus Salam donated the entire amount to the teaching of science in Pakistan.

And François was receiving more and more invitations to teach and lecture abroad. He also took part in several sessions of the Solvay Conference — we mentioned previously his first participation in 1967 (Section V.5).

He savored these invigorating get-togethers, where he felt supported by a community of physicists, a much appreciated support.

Science, his friends, his physicist friends, these would always act as a beacon for him, a guiding light whenever his personal life got a bit too rocky.

VII.3. Aharon Casher and Tel Aviv University

It was always with the greatest pleasure that François got together with his friend Aharon (Roni) Casher, whether at CERN, in Brussels, or at his home in Israel.

Aharon Casher, born in Haifa in 1941, was a professor of theoretical physics at Tel Aviv University.

François and Roni had known each other for a long time. François and his second wife Danielle had frequently spent time in Israel at Roni's, and Danielle had become fast friends with Roni's wife, Judy.

In 1981, François invited Roni Casher to come work at the ULB for two years. Ever since, they closely collaborated. Casher returned regularly to visit François at the Pool de Physique, and conversely, François was a regular visitor to Tel Aviv University. Aharon and François shared the same passion for the study of quantum gravitation and the physics of black holes. Like François, Casher was a free spirit, an impartial judge of the new ideas proliferating in physics, and like

François, he had a scathing sense of humor. He was quite critical of string theory, considered back then as the "ultimate theory of nature" by a majority of physicists. Even though they had both contributed to positing a model where all super-string theories would be contained in a single "boson string," Roni held on to his self-deprecating streak: *Very well, we'll prove it, and then we'll kill that one, and then they'll all be dead*!

François was also rather skeptical when it came to a grand unified theory that would govern all phenomena, considering himself an agnostic with regard to this supposed "Holy Grail of Science." He even envisioned that the universe might not be entirely knowable; and besides, was there only one universe?

Roni Casher was a captivating teacher who attracted many young researchers to work with him. He was also an avid reader, omnivorously curious, versed in a wide range of subjects, with a generously flamboyant side, and always ready to support a colleague's career. Like François, he was not chasing after glory, but indulged in science for the sheer pleasure.

And Casher was not one to deny himself …

He was a gourmet, a knowledgeable epicurean.

But he was sometimes a bit absent-minded, a little too casual, one who had no compunction about showing up in worn-out jeans, shirt untucked, at a three-star restaurant in Burgundy, conveniently located on the road between Brussels and Geneva, the ULB and CERN, a very familiar route to the two friends. François, dressed appropriately in a suit, tried to get him to at least wear a tie, but Casher, who had never worn that kind of thing, wouldn't let himself be "strangled."

During one of his many stays in Brussels, he was very generously housed by friends. He'd come with the firm instructions of his wife who had stayed behind in Israel: "Don't come back without chocolates, my love!" Of course, the famous Belgian chocolates … "Don't even think about returning without the precious viaticum."

Casher purchased the eleven prescribed boxes of chocolates and stowed them in his luggage. He then left his hosts without further ado! The friends were outraged and would take a long while to get over this affront: "Not a single box for us! Not one tiny piece of chocolate!"

Starting in 1984, François was a professor at the School of Physics & Astronomy at Tel Aviv University. There, he found not only a team of top-rate researchers, but a work atmosphere very much to his liking: no feverish competition, no aggressiveness, everyone full of good humor and driven by the same insatiable curiosity before nature's enigmas, determined to practice their art in a relaxed, joyful climate, and completely devoid of that "idea theft" paranoia that reigned in certain departments. All spoke freely of their research and shared their results, a deontology that François has put into practice throughout his life.

Between theoretical discussions, they played chess and chatted. For François, who carried a perpetual burden of worry despite the happiness his work provided, his time in Tel Aviv was both a relief and a stimulant.

In addition to Roni Casher, there were Yakir Aharonov and Shmuel Nussinov — the three musketeers, flanked by François, their inseparable fourth. This quartet formed the core of cutting-edge research in quantum physics.

Yakir Aharonov was a remarkable physicist and an equally excellent chess player. His wife was convinced that François and her husband would one day share the Nobel Prize. Yakir did have to his credit a very important discovery, made in 1959 with his colleague David Bohm, the Aharonov–Bohm effect.

In 1998, Yakir and Michael Berry would receive the Wolf Prize for their respective research in quantum physics.

Yakir loved to be surrounded by students who shared his taste for conceptual adventure, such as Lev Weidman, with whom Yakir strove to shed light on the phenomenon of "superposition."[13] François met with Lev one Saturday at the university and got into a discussion on the subject. Lev froze — it was Shabbat. Lev was not at all religious, and yet, he willingly complied with certain requirements out of a desire to manifest his membership in the community. Not talking shop on Saturday was one of those stipulations!

Another student was Sandu Popescu, a Romanian Jew, husband of Ildi, a Hungarian illustrator. Popescu was Yakir Aharonov's best student,

---

[13] In quantum physics, a same particle, at the same moment, can "inhabit" several different places or states.

and also worked with François. Along with Jeff Tollaksen, of Chapman University in California, they carried out bold thought experiments.

During their stays in South Carolina where they participated nearly every year in Columbia University's summer institute, the little Tel Aviv foursome excelled in keeping the restaurant dinners lively with chitchat of all sorts, and not always about physics.

Shmuel Nussinov, an equally remarkable physicist, proved brilliant at devising the most far-fetched theories which he shared at coffee breaks. He also enjoyed playing the part of the miser: in South Carolina, when everyone else was feasting at a Chinese restaurant, he ordered a one-dollar bowl of rice, only to discover that all he had on him was ninety-nine cents, to the annoyance of his fellow diners. But a few days later, he invited François and a Columbia University secretary out to dinner at a three-star restaurant.

In Tel Aviv, François also met Professor Cobi Sonnenschein, who had chosen to live in Neve Shalom/Wahat as Salam, a village unlike any other, where Jewish and Arab Israeli citizens live side by side. Founded in 1969 after the Six-Day War, located equidistant from Tel Aviv and Jerusalem, not far from Rehovot (Weizmann Institute), it had a few hundred inhabitants who were all committed to showing that it is possible to live together in harmony. Today, their village remains an educational experiment in peace-making.

It was in Neve Shalom that the physics departments of Tel Aviv University, University of Jerusalem, and the Weizmann Institute choose to organize twice-monthly joint seminars in theoretical physics.

In the 1990s, François worked with Professor Yuval Ne'eman who pursued concurrently careers in the physics of elementary particles and in politics. Founder of a right-wing party, he sat in the Knesset, and then became Minister of Energy and Infrastructure. He was a charismatic and cultured man, supported at the university by the formidable Matilda, his highly efficient secretary, an Iraqi Jew whom everyone feared, and who also had an uncanny knack for changing shekels into dollars, to François's great advantage since for a foreigner, at that time, being paid in shekels, which was not yet a negotiable currency, posed serious problems.

It was also at Tel Aviv University that François first met Leonard (Lenny) Susskind, who was a visiting professor there. A plumber's apprentice at 16, Lenny managed to get into City College New York and would later teach at Yeshiva University. In 1979, he was appointed as professor at Stanford. Lenny was a character in the world of physics, considered one of the fathers, along with Yoichiro Nambu, of String Theory. He was also one of the architects of the "multiverse" or "cosmic landscape," a theory that posits the existence of multiple universes, and of so-called Technicolor theories, theories where the scalar condensates (Section V.3) are replaced by more complex processes. Once at a summer institute at the University of Colorado at Boulder, Lenny, François, and Robert Brout were all in attendance, and together they toured the region.

During one of François's stays at Stanford, on Lenny's invitation, Mira and Lenny's wife would become friends.

VII.4. The Florida Crowd

In the 2000s, François took part in the annual conferences held by the University of Miami successively at Coral Gables, Key Biscayne, and Fort Lauderdale. Given the local climate, the "summer institutes" were held in December, beneath a cloudless, sunny sky in a tropical paradise: sandy beaches, palm trees, luxury hotels, swimming pools, and fine restaurants. Unsurprisingly, these "winter institutes" were hugely successful and a large number of physicists participated faithfully every year.

This was how a group of vacation-loving regulars, who also enjoyed the work sessions, came to be created. François felt like a fish in water among these merry conference-goers.

There was Lars Brink of the University of Göteborg whom François already knew from the time Lars invited him to Sweden to play the "prosecuting attorney" at a dissertation defense. There were also Thomas Curtright, of the University of Miami, and his charming wife Joann, Charles Thorn, excellent tango dancer, and his wife Mary, the Franco-American Pierre Ramond, both of the University of Florida, Kelly Stelle of Imperial College London, a longtime friend François first met at CERN, and Sasha Polyakov, eminent Soviet physicist from the Landau Institute in Moscow, later a professor at Princeton.

Kelly Stelle, François, and Mira organized a fabulous excursion to the Florida Keys archipelago, from the Key Largo of Humphrey Bogart fame all the way down to Key West, island of the bohemian intelligentsia, haunted by the ghosts of Ernest Hemingway and Tennessee Williams.

Pierre Ramond held Mira in high esteem. Whenever he met up with François, he would ask: *Hi! So, where is your better half?*

As for Sasha Polyakov, François fondly recalls the times they were together. In 2005, the Solvay Conference on Cosmology was held in Brussels, attended among others by Steven Weinberg, Eliezer Rabinovici, Gerard't Hooft, Sasha Polyakov, and Andreï Linde, from Stanford. Andreï Linde gave a paper on inflation and the anthropic principle in cosmology, a principle which, according to some, tells us that "things exist as they are because we can observe them." François was not an enthusiast of this principle, finding it analogous to Darwin's natural selection.

During the discussion period that followed, François took the floor:

I would like to make some comments that express my unease when it comes to the way the anthropic principle is used here. It inevitably calls to mind the principle of natural selection in biology. Well, I don't want to discuss natural selection in biology but I think that transposing a principle from biology to physics is somewhat dangerous. A simple example: let's suppose that we don't know the theory of gravity and let's suppose that someone wonders why apples fall to the ground (we know the historical importance of apples in gravitation). The response is extremely simple: if they didn't fall to the ground, apples would not generate trees, which is why apples that do not fall to the ground have disappeared. Of course, we would like to say that we should never reason so simplistically. And that we had better get busy and find a decent theory of gravitation.

(We can take delight in the fact that one day an apple did have the great idea of falling on Newton's head.)

A while later, François came across Sasha Polyakov, who declared in a burst of laughter: "François, you made my day!"

This was the same Polyakov who, when asked before his departure from the USSR for the United States what he thought about China, responded, "It's nice to think that there are places that are worse than yours."

What François particularly enjoyed about these festive Floridian get-togethers were the after-dinner talks that crowned the banquet celebrating the annual reunion.

Put in charge of this task one day, François chose a comic approach to the two trends present throughout the development of Western physics:

- Greek thought, with its harmonious, aesthetic, geometric vision of the universe, its dominant ideal of symmetry symbolized by the sphere and the Platonic solids, illustrated by ancient Greece's mathematics of the cosmos (Section III.1)

versus

- Jewish thought, where the answers to the questions humans ask about the universe are found in the subtlety of decrypting its complexities.

In Jewish thought, if symmetry is there, it is only by breaking it that the universe can develop, as exhibited in the Tsimtsum[14] theory of the Cretan astrologist and mathematician Joseph Solomon Del Medigo, trained in Padua under the tutelage of Galileo, and later rabbi, music theorist, explorer, and philosopher: *one day, God, the One and Almighty, decides to create the universe, but God is everywhere, so there is no room for the universe! So God started by shrinking himself down to a single point, and then fills the newly available space by dilating into it, but this gives rise to an isotropic, homogenous universe, impossible to differentiate or organize, so that didn't work ! So, back to the drawing board, but this time, as he contracted, he leaves here and there in space little bits, "residues" ... And that is how differentiation started and a veritable universe was created.* Del Medigo summarized this vision with an eminently prescient phrase:

"It is only by breaking up perfect forms that they can reach their perfection."[15]

Thus was broken the primordial symmetry!

---

[14] The Hebrew word meaning *contraction*; the concept in the Kabbala dealing with the process preceding the creation of the universe.

[15] *Elim*, published in Hebrew in 1629 in Amsterdam.

VII.5. Prolegomena to the Quantization of General Relativity

Quantum physics enables us to understand how different components of matter interact.

General relativity does the same for our understanding of space and time. It has given us the revolutionary notion that space-time reacts to the presence of matter. It bends and warps ... it has become a dynamic physical object, a field like any other, in a way. For example, the expansion of the universe means that space-time is dilating.

These two theories function nicely, they are acknowledged by all and each has been proven by countless experiments.

And yet, as we have noted several times already, they are incompatible with each other!

We shall return in the Epilogue to this notion of incompatibility.

For François, two incompatible theories cannot both be correct. There cannot be two separate truths. If we want someday to understand the origin of the universe, we must figure out the quantum character of gravitation. And perhaps undertake an in-depth revision of our current concepts in physics.

Starting in the 1980s, François would take to this enterprise with gusto, a task in synch with his penchant for subversion. String theory, superstrings, supergravity, supersymmetry, fractals, 7-spheres, emergent fermions, infinite algebras, concepts of space and time — no stone is left unturned. For over twenty-five years, François, often alone or with fellow travelers Philippe Spindel, Aharon Casher, Anne Taormina, Jean-Marie Frère, Yakir Aharonov, Jean Orloff, Laurent Houart, Paul Windey, Hermann Nicolai, Auttakit Chattaraputi, Marianne Rooman, and Peter West, would explore each hypothesis, mobilizing all the resources of his intuition, his imagination, and his enormous mathematical knowhow. He would become a link in the chain of physicists that took on that great challenge: to reconcile general relativity and quantum physics, physics of the cosmos and physics of the infinitely small, for a theory that would unify quantum physics and general relativity would open the door to finally understanding phenomena involving huge amounts of matter or energy on "infinitely small" spatial dimensions, among which, for example, is the genesis of the universe, or black holes, a phenomenon that ranks among those curiosities that spur researchers in their quest to reconcile quantum physics and general relativity.[16]

François tackled black holes in 1993. He published several articles, alone or in collaboration with Laurent Houart, Paul Windey, Philippe Spindel and Aharon Casher.

And this was but the tip of the iceberg.

A shimmering palette

We have been presenting in this episode the main topics dealt with by François since 1964, in chronological order according to his publications. This fails to do justice, however, to the continual profusion of his interest areas and his astonishing capacity to juggle several things at once, though never treating any one area superficially. The complete list of his publications calls to mind a palette containing the whole range of physics' colors, with dabs of the most cutting-edge mathematics, like this study, for example, conducted with K. Peeters and Anne Taormina in 2008, on icosahedral viruses.[17]

Robert Brout was best able to describe the fruitful career of his "friend of four decades" in his keynote speech entitled "François Englert, friend, colleague, man of science and humanist" at the conference held in 1999 at the ULB to celebrate François's accession to emeritus status:

> *As I have been looking back on François's career (the Englert saga), I realize that in fact, I've been tracking a good part of the development of fundamental physics over these past few decades. You will forgive me then if I am only able to skim over the successive stages of François's*

---

[16] Recall that a black hole is a celestial object so dense that the intensity of its gravitational field prevents any form of matter or radiation to escape. It is believed to have been formed by the gravitational collapse of a massive celestial object.

[17] Englert, F., Peeters, K., and Taormina, A. "Twenty-four Near-Instabilities of Caspar-Klug Viruses," in *Physical Review E,* vol. 78, October 2008.

An icosahedron is a fascinating geometric solid, one of the five platonic solids, and equally fascinating is the capsid of the virus (their protein shell) whose structure is mostly very regular, either icosahedral or helicoidal (like the infamous COVID-19).

accomplishments, a summary that will not even come close to the atten-tion they deserve.

  [...]

  The universe of François Englert, his contributions to physics and the wealth of his collaborations with so many people, most of whom have become his closest friends, have compelled me to perhaps go on a bit too long. But that is the man he is. And such is his body of work. So many have been inspired by his example, his scientific clear-sightedness, and above all the humanity that imbues his contact with others. I hope I have succeeded in conveying in this address a bit of François's vitality and depth, man of science and humanist.

Apart from his numerous publications in scientific journals, François had also contributed to reference works, among which are the following:

"From quantum cosmology to quantum gravity" in *Unification of the fundamental particle interactions,* 1983.

"The quest for unification" in Recent Developments in *Quantum Field Theory,* 1985.

"Cosmologie quantique" in *La nouvelle encyclopédie Diderot, Aux con-fins de l'Univers,* 1987. (Englert, Robert Brout).

"From Quantum Correlations to Time and Entropy" in *The Gardener of Eden* (Festschrift for Robert Brout's 60th birthday) 1990.

"La notion de temps" in *Le vieillissement,* Laus Medicinae, ULB, (Englert, Brout, Aharonov) 1997.

"Spontaneous Compactification of Eleven-dimension Supergravity" (Englert) and "Gauge N=8 Supergravity and its Breaking from Spontaneous Compactification" (Englert, de Wit, Hermann Nicolai and Biran) in *Modern Kaluza-Klein Theories* 1983, volume of the collection *Frontiers in Physics.*

With regard to these latter contributions, an anecdote is worth repeating:

François and the guru

The Kaluza–Klein theories, conceived in 1919 by Theodor Kaluza, give a physical interpretation to certain geometric properties of compact spaces.

These theories, which introduce notions such as supergravity in eleven dimensions and the 7-sphere, are likely to intrigue the uninitiated and might sound utterly magical.

The fact remains that these notions one day struck the imagination of the famous Indian guru Maharishi Mahesh Yogi, founder of the Transcendental Meditation (TM) movement, guru of the Beatles, the fabulously wealthy pope of American counterculture in the 1970s, and incidentally a graduate of the University of Allahabad with a degree in physics.

In January 1958, he undertook the first of several world tours in order to *bring to everyone, everywhere in the world, the experience of pure consciousness through Transcendental Meditation.* He created dozens of meditation centers in Southeast Asia, the United States, and Europe, as well as at colleges and universities. He was estimated to have initiated nearly six million disciples to TM. A significant figure, in other words, but as we can read in the Economist obituary, on February 2008, "Crank? Crackpot? Charlatan? Maybe all three. Yet the maharishi was generally benign."

The Vedas[18] were the cornerstone of his movement, where alternating between Vedic science and modern science was at the heart of his doctrine. For instance, starting in the 1970s, he enjoyed embellishing his speeches with such terms as "phase transition, superstring theory, supergravity, unified field ..." which he then connected almost whimsically to *a unified transcendental field identical to pure Vedic consciousness.* Out of this emerged a whole *quantum mysticism.* The work of Bernard de Wit and Hermann Nicolai on the generalization of supergravity, introducing "gauge $N = 8$ supergravity" to the tune of enormous equations,[19] could only have enchanted the famous guru. He invited de Wit and Nicolai to participate in a seminar cruise on the Rhine during which the famous equations were projected onto a screen.

In 1982, François, Bernard, and Hermann recalled the guru's wild fascination one time during a stay in Amsterdam, where they were in the process of writing their paper entitled *Gauge $N = 8$ Supergravity and its*

---

[18] Sacred texts revealed to Hindu sages dating back probably to the 15th century BCE.

[19] de Wit, B. and Nicolai, H. "$N - 8$ Supergravity and Local SO(8)xSU(8) Invariance," in *Physics Letters,* vol. 108B, 1981.

Breaking from Spontaneous Compactification. They decided, as a kind of joke, to "pay tribute" to the Maharishi. In the conclusions of this otherwise perfectly serious paper, there appeared a sentence concocted by the three friends who by then were in hysterics: ... *in this way, gravity is indeed decoupled of all scalar excitations, and thereby, states of higher consciousness are being promoted to the ultimate vacuum.* A sentence whose second clause is strictly meaningless.

The paper was published in 1983 in *Physics Letters*[20] without any editor or outside reader noticing a single thing!

Likewise, a bit later when, upon the request of Peter Freund who was working in Chicago with Yoichiro Nambu, the article would be reprinted in the volume *Modern Kaluza-Klein Theories* in the "Frontiers in Physics" collection, and again, no one saw through it!

Oversight, tacit complicity, sense of intimidation at the high standard that the paper represented? No one will ever know ...

Intermezzo: Farewell to the parents

We last saw Joseph and Baltche in the immediate post-war period. What was their life like since then?

Once back in Brussels, they resumed their wholesale fabric operation, where business was soon back to its normal pace.

With the departure of Marc, who left home to go live with his partner Anne Chwoles, François was thinking about moving upstairs into one of the garret rooms, where he would have more personal freedom, but without having to really leave home.

Baltche was delighted to have her youngest still under the same roof. She hoped to keep him as long as possible, in fact. So when François announced his intention to go live with Esther, her heart sank. Even though Esther is their friends' daughter, this marriage saddened Baltche.

In the early sixties, the parents left the noisy Boulevard d'Ypres and purchased an apartment in a quieter neighborhood, Quai aux Pierres de

---

[20] Biran, B., de Wit, B., Englert, F., and Nicolai, H. Gauge $N = 8$ Supergravity and its Breaking from Spontaneous Compactification, in *Physics Letters,* vol. 124B, 1983.

Taille, a short walk from the very "Flemish Renaissance Revival" KVS, the Royal Flemish Theater.

But Baltche soon started having circulation issues and had a heart attack. She recovered, though she remained fragile.

Hoping to avoid the harsh Belgian winters, François's parents decided to rent a pied-à-terre in southern France, in Menton, where they could stay during winter. Menton is a charming seaside resort (Élisée Reclus called it the pearl of France), with its particularly pleasant micro-climate, spas, and a casino. But winters there can be windy!

During one of François's visits to Menton, his father explained the plan to him: when they went out for a walk, at every street corner, Joseph acted as a scout to make sure Baltche didn't get caught off guard by a wind gust.

During François' visits, his mother displayed all the Yiddish mama's pleasures laden with anxiety: worried at the idea that her dear youngest, now almost forty, might fall off the cot he was using for a bed, she barricaded him with a chair!

Likewise, when she learnt that François and Esther were not getting on very well, she sent her husband as ambassador, though the father's embarrassed intervention did little to improve the situation.

In the late sixties, having decided to leave the business, the parents were considering leaving the Quai aux Pierres de Taille, especially since an incident caused by a pigeon made the apartment abhorrent in their eyes.

Like many Jews of their generation, the Englert parents had little sympathy for animals, domestic or otherwise. One day, a pigeon flew in through a window, and started fluttering around the room in a panic. Equally panicked, Baltche called the fire brigade who, poorly trained in pigeon trapping, made a real mess of things. The couple since wanted nothing more than to leave the scene of the crime for good.

The next move was to Avenue d'Italie, in Ixelles, where they purchased a pleasant little apartment.

When François and Danielle's marriage appeared to be floundering, a very depressed François occasionally squatted in that apartment during his parents' visits to Menton.

In 1973, Baltche suffered an embolism. She was rushed to the hospital, still conscious. When she saw her husband and two sons at her bedside, she realized exactly what was happening. With a hint of self-derision, she sighed,

Oh good, everyone is here! That must mean it's not looking good for me …

The next day, when François returned to her bedside, he found her in a semi-conscious state, moaning in pain. Outraged, he called for the doctor to come administer some morphine. Baltche would leave this world in her sleep, peacefully.

Such was the death of the mother who François had always felt to be rather distant, but who probably was not so at all. Like so many Jewish mothers (and many other mothers, actually), total devotion to the family — a concern for doing the right thing, the children, and perpetual anxiety in the face of real or imaginary evils that might descend upon all of them — doesn't leave them much latitude for expressing the full extent of their love. The eternal paradox, further exacerbated in a woman who carried buried within her heart a grief for not only a whole family but also a mourning of a world forever vanished. Baltche was one of those women who remained prisoner of a deafening silence, because some things always remained unspeakable.

Joseph would live a few years alone on Avenue d'Italie, getting by, helped by a housekeeper, until a female friend of the family took it upon herself to find him a new spouse! On the advice of God only knows which *shadchanit* (matchmaker), she had a woman brought over from Israel, a perfectly honorable lady whom she introduced to Joseph. Joseph was not thrilled at the idea, but a few weeks later, he could no longer bear his "match's" exasperatingly incessant gossiping. The family friend intervened once again, and the woman disappeared from Joseph's life for good. He was relieved, but this sudden departure also felt like a failure for Joseph. The tragic side of his life was all too apparent and weighed upon him, such that he soon fell into a deep depression.

François and Marc often went to visit. They both got along with their father. François had always felt closer to his father than to his mother, and Joseph, in his twilight days and despite his introverted nature, talked openly with his younger son.

As time passed, Joseph and François came to resemble each other more and more, physically but also by certain character features that started to quietly express themselves in gestures of affection.

On the advice of François and Marc, Joseph decided to move into a Jewish assisted living facility, the *Heureux Séjour*, in Saint-Gilles. Marc devotedly oversaw every aspect of the move.

It was the right decision. It was in this rest home that he would live out his days, feeling somehow revitalized and able to overcome his depression. Though he did enjoy discussing the Torah with a few pious elders, he quickly acquired a reputation as a fun-loving partner to the ladies, who greatly enjoyed his company: "He's a real mensch," they said to François. "In fact, he's the only man in this place!" Joseph was over the moon! So much so that once when François stopped by for a visit, he seemed vaguely chagrined to have to get up and leave his circle of female admirers. But this feeling was quickly dispelled, for it goes without saying that Joseph was always happy to see his son, and was also very proud of him, delighted to introduce him to the other residents.

You see, how proud I am of you, but you can also be proud of your old dad!

François took part in the Passover Seder at the *Heureux Séjour*. Joseph was especially delighted when his son brought along a female companion, sensitive as he was to the beauty of young women. He made sure to bring them all sorts of traditional macaroons and treats associated with this Jewish holiday. Joseph began recounting in minute detail the events of his life at *Heureux Séjour*. Certain residents were particularly interesting, such as that elderly Jewish man who used to know Sakharov.

Joseph would die of a stroke in September 1987, at the age of ninety-two. For the few days leading up to his death, he seemed to be overcoming the attack. At any rate, his humor was unaffected.

He exited this world, a world that had not always been kind to him or his family, with a final quip.

When the nurse came by to adjust his pillow, she asked,
Is your pillow alright like this?
He replied with a sigh,
The pillow's fine, I'm the one who's not alright.
The end was near. Joseph fell into a partial coma.

When François, having rushed to his bedside, leaned down close, he momentarily regained consciousness. A final gaze.

As François recalls this episode for me, I sense all the buried nostalgia and regrets resurfacing.

Regret at not knowing how to better communicate back when they were all living together. Regret at not trying to get to know his father better, a man at once discreet, intelligent, and surprising.

A dominating, lively, resourceful, and pragmatic mother, a father more calm, tending toward the more thoughtful, more mysterious, this is the breeding ground that gave rise to François.

Does this explain that? Does this explain the man we are attempting to grasp?

And what has been the role of his lived experience, so diverse, in such changing times, such singular circumstances?

Have we added any pieces to the puzzle?

Undoubtedly.

Even though we know,

It's much more subtle than that.

Episode VIII

Israel and the *Jerusalem Impromptu*

VIII.1. At Joseph Katz's

Among the many countries where François went to teach, take part in seminars, or meet with colleagues and friends, Israel holds a special place.

It was in 1968 that François made his first trip there.

He was part of a delegation of the International Committee for a Negotiated Peace Agreement in the Near East, created in 1967 after the Six-Day War, at the initiative of Marek Halter (Section II.1). This was an official invitation by the Israeli government. The delegation was made up of, among others, Marek, his wife Clara, Pierre Verstraeten, and Jean-François Revel. The goal was to bring parties closer to a negotiated peace, those parties being the Israeli intellectuals and leaders, on the one hand, and the Arab intellectuals and leaders on the other hand. They met David Ben-Gurion, founder of the Israeli state, at his home on the kibbutz Sde Boker in the Negev desert, an intransigent, no-nonsense man who responded unhesitatingly to their questions. Escorted by two military jeeps and a handful of soldiers, they drove to the newly conquered West Bank. On the way, they heard the sound of an explosion, a bomb attack had just taken place not far off, and they witnessed the strong-armed arrest of some Arabs. All along the way toward Hebron, the atmosphere was extremely tense. In Hebron, they met with the Arab mayor, who received them and spoke through a young interpreter. The discussion had hardly ended when the mayor turned to Marek Halter, whom he seemed to know, and addressed him amicably in French! Next, they went to the Gaza Strip, again under army escort. They marveled at the sumptuous landscape, the miles of beach along the Mediterranean. But Pierre Verstraeten and François decided to break from the group and walk to Gaza City, where they discovered a miserable, dilapidated town. They were invited to have tea with a Gazan they met, just to talk. They learnt a thousand interesting things, and also that many desperate Gazans wished they could leave Gaza, which was why people there were constantly asking them for their addresses and contact information in Europe. They visited schools built by the Egyptians, where they discovered drawings on the wall representing gas chambers where Jews were exterminated. One wondered what the day's lesson was about that would give rise to such drawings. François, puzzled, could only guess. When the two friends rejoined their party, they noticed that their little side trip had considerably vexed their escort!

In the end, the Committee's action had little effect on the inextricable political situation that persists in the Middle East even today.

We already know that François was a regular visiting professor at the University of Tel Aviv and at the Weizmann Institute in Rehovot. Later, he would be named "Senior Professor by special appointment" at the University of Tel Aviv.

During his frequent stays, François was hosted in Jerusalem by his friend Joseph Katz, who had settled down somewhat since the earlier days of their friendship — no longer the firebrand physicist in threadbare clothes and worn-out sneakers pacing the halls of the ULB, in constant conflict with the university administration; nor the model husband devastated by the failure of his high-society marriage, following which he decided to "live a life of unbridled passion."

Disheartened once and for all by what seemed to be his unavoidable fate in Belgium, Joseph Katz decided in 1968 to "set sail." Once in Israel, he got married, became a father, and in 1980 was appointed professor at the Racah Institute of Physics at Hebrew University in Jerusalem.

He was a longtime friend of François, with whom he shared many things. Joseph was also a "child of the Shoah": with a father murdered by the Nazis, a deeply traumatized mother, and a succession of adoptive families that never managed to tame this angry child. Both were disenchanted with their training at the Polytechnique and both chose the path of physics. For François, every visit to Israel was a welcome opportunity to get back together with Joseph Katz and especially to observe that his turbulent fellow physicist had found in Israel a certain inner peace. He married Ruthi, a powerful woman, instructor in the Israeli army, with whom he had two adorable children. He who, back in Belgium, seemed the perfect illustration of Gide's famous pronouncement, *Familles, je vous hais*, a declaration of hatred for families in general, took fatherhood very much to heart. Every day at noon, he came home to make lunch for the children, supposedly because "Israeli women don't know how to cook." Joseph had become a kind of *yiddishe-tate*, a father hen, with all the culinary zeal that came with it. Whether it was for his offspring or his friends, it was a proven fact, he was a divine cook.

But during his times in Israel, François had other encounters, much more surrealist ones.

He came across Harry F. there, who, while pursuing his very serious research in the field of nonlinear optics and perhaps because of his immersion in the very religious environment of Bar Ilan University, was also studying the Kabbala, getting deep into the most obscure esoterica. He seemed to be wrestling with serious psychological problems and sought answers in the study of paranormal phenomena.

He put François in contact with one of his psychic friends, Miri, who had the privilege of speaking daily with God. One day, Harry announced to François, "Something truly catastrophic has happened, God has taken ill!" He mandated François to get Miri to find a solution to this unprecedented disaster. By way of answer, Miri and François slipped away together to England where Miri was taking part in a convention at the famous College for the Advancement of Psychic Science. In London, François ran into his friend Kelly Stelle who invited the couple over. Kelly, son of protestant missionaries in China, was very impressed by the beautiful and sophisticated Miri.

But the long conversations that Miri claimed to be having with the dead only exasperated François, in the end, and he decided to break it off.

This brief frolic would serve as a prelude to a few other flings, just as brief, with equally eccentric Israeli women. There would be further trips to England, introductions into the world of theater, with a few great Shakespearian actors as a bonus.

In 1993, while staying with Katz, François met Alexander (Sasha) Abramovitch Belavin and his wife Ira, from the Russian Federation, the former Soviet Union.

Despite the shortages and hassles of the Soviet regime, despite the less than inspiring regimes that succeeded it, the Belavins remained much attached to their country. Of the Orthodox faith, they were people who exuded kindness and generosity. Under the Brezhnev regime, for example, Sasha had organized a more or less underground institution of higher education for Jewish students left out of university by a quota system for Jews, evidence of the prevailing anti-Semitism. His school would be acknowledged officially at a later date.

Even before the fall of the Soviet regime, Sasha was a celebrated physicist in Russia, hailed throughout the world, and particularly in the United States and in Israel where he regularly traveled to lecture and to complement the pitiful salary he was earning in Russia. In 1998, François met up with him again during a sabbatical that he was spending at Hebrew University in Jerusalem with Eliezer Rabinovici. Every time he got together with François and Mira in Israel, Belavin always claimed he had come to Israel "only for the pleasure of seeing them again." But this was also the time of Michael's wedding, Joseph Katz's eldest son. Since neither Sasha nor Ira had the "right" clothes for a chic ceremony, they got dressed head to toe from what they found in François's and Mira's wardrobe. Later, François and Mira would go visit them in Russia. Mira and Ira, who ran a children's choir, would become great friends, despite the fact that Ira spoke only Russian — but luckily a Ukrainian Russian whose Polish inflections Mira could recognize, since Polish was her native tongue. Sasha and Ira lived in a small, run-down apartment that belonged to the famous Landau Institute, the "ghost institution" where Sasha had been working for a number of years. This was the apartment where the eminent physicist Sasha Polyakov lived prior to his departure for the United States. A "ghost institution," for under the Soviet regime, when it came to theoretical physics, it had no material existence, no building. The physicists worked from home, which proved advantageous in the end, since they were able to avoid administrative oversight. This was also one of the rare Soviet institutes still run by a Jew, Isaak Khalatnikov, a disciple of the great physicist Lev Landau.

In Russia, Mira had the wonderful surprise of finding people from her father's family, and among them, linguists and a number of physicists! François is unlikely to forget the frantic race by one of Mira's cousins to find in a village near Moscow an accordion to give to one of his daughters as a present — a physicist cousin who, in the course of a rather delirious phone conversation with Sasha Belavin about accordions, discovered that he was actually talking with his professor from the Landau Institute. On every trip Sasha and Ira made outside Russia, the friends were eager to get together.

Israel would always remain a meeting place.

We have already stressed the importance of the universities of Tel Aviv and Jerusalem (VII).

The Katz apartment was also an enduring hub for friends to gather, sometimes producing surprises:

For instance, one day, François was there when, along with her boyfriend, a woman showed up who turned out to be Nicole, the daughter of his good friend Robert De Mayer! Joseph Katz, way back then, had often played babysitter for Nicole, who felt great affection for him, and here she was coming by to say hello while on a visit to Israel. A fun reunion for all concerned!

When François was awarded the Wolf Prize (with Robert Brout and Peter Higgs) in 2004 in Jerusalem, Jean-Marie Bertrand, François's old roving partner, decided to come along. Jean-Marie, ever resourceful, found a nice homestay arrangement, while François and Mira, as always, stayed with Joseph Katz.

And there, Jean-Marie disclosed an unexpected side to his rugged loner personality: for entire evenings, he recited poems with much skill and emotion. He also improvised goofy odes on François and his Wolf Prize, spoofing Gerard de Nerval or José-Maria de Heredia.

> *Ils allaient conquérir le fabuleux … Prix Wolf*
> *Que Cipango mûrit dans ses mines lointaines,*
> *Et les vents alizés inclinaient leurs antennes*
> *Aux bords mystérieux du monde Occidental…*[1]

---

[1] "They set sail to conquer the fabulous … Wolf Prize/That Cipango produced in distant mines/. As trade winds stretched their tentacles/Over the mysterious rim of the Western world." In "Les conquérants de l'or" published in *Les Trophées*, Paris, Lemerre, 1893.

His audience was spellbound, to the great delight of Joseph Katz.

One of these Katzian encounters would have a lasting impact on François's life.

On one of his work trips, François stayed as usual at Katz's. Weary of his mixed fortunes, both back home *au plat-pays*[2] and in the land of Israel, he declared himself to be "on a sex strike"!

But in fact, François was not alright at all. Although he had kept his friendship with Robert intact while broadening his field of research, thereby forming new collaborations that widened his circle of physicist friends internationally, he had yet to feel secure affectively. And he knew that without this haven, he could not extinguish that recurring anxiety of the lonely, abandoned child in a deeply hostile world.

Ruthi perceived François's distress, she who believed so strongly in the beneficial effects of family. With that in mind, one evening she invited one of her young women friends to come "grab a bite" with François, *casser la graine*, as Joseph liked to say. There was immediate chemistry between them. Joseph declared, somewhat peremptorily, "The sex strike is over!"

This was in 1990, and the young woman's name was Mira Nikomarow.

VIII.2. Mira

Few people know Mira's "pre-François" life, before she landed unannounced in Belgium, one fine day in August 1990.

Mira comes from a Russian-Polish Jewish family, her mother Judith was born in Warsaw, her father David in Byelorussia, in Vitebsk, the town of painter Marc Chagall.

In 1939, Judith was 20.

The invasion of Poland by German troops in 1939 would radically and tragically disrupt life in that country.

Shortly before the start of the war, the violinist husband of her sister Bluma was invited for a concert tour in Minsk, the capital of Byelorussia. He suggested that his pregnant wife come along, as a way of dodging

---

[2] *Le plat pays*, or low country, is a song made popular by Jacques Brel, an ode to his native Belgium, known for its flat landscape.

the German troops. But it was only after the terrible bombing of Warsaw, on 1 September 1939, that she left for Minsk along with her sister Judith. Right before their arrival in Minsk, the border between Poland and Byelorussia was suddenly shut down. The two sisters found themselves on the Russian side, and Judith would remain there throughout the whole war.

The Jewish community of Minsk did what they could to help the refugees. Judith found work in Vitebsk in aerial military photography. It was there that she met her future husband, David Nikomarow, officer engineer in the Red Army. David had three sisters and a slew of nieces and nephews. With Judith, who felt terribly isolated in Vitebsk, he shared his large extended family with whom she would forge strong bonds. She quickly picked up Russian and considered her husband's family as her own.

Some Nikomarows had immigrated to the United States in the 1920s, though they did not renounce the Communist ideals of their generation. This is how Mira would be able, during her trips with François, to meet up with so many cousins, whose leanings were and remained staunchly toward the left. One cousin even became mayor of Ithaca, one of the rare towns in New York State that had always voted Democrat, and the place where François lived during his first stay in the United States (Section IV.2)!

But the German troops marched on the USSR. David had to leave for the front. During a bombing of Minsk, Bluma, her husband, and their child got killed in the hospital where the child had just been born. The women of the family then decided to retreat to Moscow, a smart move, as it turned out, for in the kind of airlock created in Minsk between invaded Poland and Russia, many Jews became trapped. They would all be ruthlessly massacred. During the three years of German occupation in Byelorussia, more than three million persons were assassinated by the Nazis, one million of whom were Jewish.

Judith decided to join her husband, stationed near the front. She found work in the region, in a kolkhoz,[3] a management position that she enjoyed.

When rumors started circulating that the Germans were nearing Moscow, the whole family left for Uzbekistan.

---

[3] A Soviet agricultural system whereby the land and means of production were pooled.

Seriously wounded at the front, evacuated to a hospital and eventually discharged, David caught up with the family in Uzbekistan. During the war, this small Soviet Socialist Republic generously welcomed hundreds of thousands of Soviet refugees, among them many Jews.

In 1943, Judith and David had their first child, a boy they named Grisha Hirsh, one Russian name, and one Yiddish that means "deer."

After the war, David's sisters, nieces, and nephews returned to Moscow. David, however, did not. Having refused to join the Communist Party and fearing the ever-present anti-Semitism in the USSR, he chose Poland with his wife and child. The refugees returning to the country were regrouped in Lubawka, a small town in Silesia, where Judith felt very out of her element. But there was no longer any question of returning to Warsaw, for she knew that her entire family has disappeared.

It was in Lubawka that Mira would be born, in 1946.

Her father David was once again hospitalized. A request for help sent to the émigré family in the United States went unanswered. It was the start of the Cold War, and it was not unlikely that this letter issuing from behind the Iron Curtain had been "lost."

David died in 1947. Mira was only seven months old and thus never really knew her father. Her mother worked as a schoolteacher in a Jewish school organized by the "Joint."[4]

In both pre- and post-war Poland, anti-Semitism was alive and well. Jews were openly insulted in the street. Mira was living in constant fear. Neither she nor her mother felt at home in their "native country."

In 1956, Wladyslaw Gomulka became leader of the People's Republic of Poland. He did not conceal his anti-Semitism, and sought to remove Jews from national institutions and from all levels of the Communist Party. In other words, he wished to be rid of the Jews, and fostered their emigration to Israel.

The message was clear, and in 1957, the family emigrated from Poland to Israel.

It was in Israel that Mira, now almost eleven, would at last find her roots.

---

[4]The American Jewish Joint Distribution Committee is the largest Jewish humanitarian relief organization, founded in 1914, and present in over 70 countries.

Mother and daughter had a deep connection. Judith told her daughter all about Poland, Warsaw, the war, their successive exiles, and their mourning. Mira understood how courageous her mother must have been, alone with two children, to face that return to a hostile country. She identified with her mother and forged a character imbued with boldness, faith in the future against all odds, and developed a determination to be independent that would consolidate over the years, much to the approval of her mother. The older brother, whose name was now Tsvi ("deer" in Hebrew), was a daydreamer with a vivid imagination, an artistic temperament matched by a warm personality. He lived somewhat in his own world but also meant a great deal to Mira. Her mother and her brother were her whole family.

The trio found a place to live in a rather rudimentary village built near Tel Aviv for new arrivals. There was no school in the village. Mira and Tsvi were looked after by the Jewish Agency[5] which sent each of them to a kibbutz[6] where they could continue their schooling.

Mira had a lot of trouble adjusting. She didn't know a word of Hebrew, and the kibbutz was nothing like anything she had experienced in her life thus far, either with her family or in school. The codes of collective life were unknown to her. She missed her mother, and her brother too. But later, once she was in secondary school, she started seeing the kibbutz as forming a genuine family. She enjoyed learning, the activities interested her, she made lots of friends, and in the end, began to fit it.

Then came military service.

In 1967, after the Six-Day War, Mira married Shmuel, a Sabra,[7] born on a kibbutz into a family with German ancestry.

Mira and Shmuel would have four children, all born on the kibbutz.

Mira experienced another very peculiar aspect of kibbutz life: the children belonged to the kibbutz. All decisions concerning them were made collectively, and parents were not to argue. Mira felt dispossessed of the most intimate part of her life, and she lost faith in the system.

---

[5] Zionist organization founded in 1929 to provide relief for immigrants.

[6] A type of collectivist village created for the first time in 1909 in Palestine by Jewish émigrés of the Zionist and socialist persuasion.

[7] Hebrew word meaning "cactus." Jews born in Israel are called *sabras*.

She realized that no personal fulfillment will be possible within the collectivity, where everyone depended on everyone else, where each person had to feel responsible for everyone else, though they may not know them very well. In this system where everybody took care of everybody else, the normal reaction was one of self-protection where one built a "standard" façade, which only served to deepen mutual misunderstandings and made it hard for a person to find his or her "place."

For Mira, it was a real disenchantment, and the beginning of a deep rift between the spouses, for Shmuel was a stalwart of the system.

In 1985, after much hesitation, she asked for a divorce and prepared to leave the kibbutz. The children would stay behind with their father.

A wrenching experience for Mira, where her overwhelming feelings of guilt were aggravated by the general opprobrium that greeted her decision.

She stayed a while in Haifa with her brother. She traveled, but came back regularly to see her children and tend to their needs. Tsvi had become an architect, and taught in several architecture schools in Haifa and Tel Aviv. He was a much loved professor, original and inventive.

In 1987, Mira moved in with a woman friend in Jerusalem, also an ex-kibbutznik, who offered her a job in her fashion boutique.

Later, she found a job better suited to her tastes in an art gallery in Jerusalem, run by Ruthi Katz, who was very impressed by Mira's artistic sensibility.

One day, Ruthi invited her over for dinner. There, Mira met Ruthi's husband, none other than Joseph Katz, and the other guest that evening, an old friend of Joseph's, none other than François Englert!

They clicked, that much was clear.

After François's return to Belgium, he began corresponding regularly with Mira.

Soon, François returned to Israel, and they met up again.

Things picked up speed after that.

On 20 August 1990, François went to meet Mira at the Charles de Gaulle airport in Paris, and they drove together to Belgium, Mira on a tourist visa.

The couple drove to Chiny where Hélène, François's younger daughter, was spending her vacation with a friend. For Mira, it was a complete disorientation. She had just left a country seared by the sun, to find herself in a house in the Ardennes in the pouring rain, surrounded by luxuriant greenery.

François had failed to notify the two girls of their arrival, which created stress and tension, while Mira felt completely lost, not understanding a word of French. But the atmosphere quickly relaxed and the evening ended in bursts of laughter from "the girls."

Mira was next introduced to the famous Uccle apartment. Its prevailing atmosphere of agitation might have surprised some, but it would take more than that to surprise a former kibbutznik!

Between Mira and François, who hardly knew each other, there blossomed a relationship of great trust. He who had always had such a hard time talking about his childhood could now open up to his new partner.

For the first month, François spoke to me at length about his war, I told him about my family, I discovered that we had so much to share, and we sensed we'd have a long-lasting connection.

François once again felt he was on solid ground — his ship had reached its harbour. And soon, with Mira and their respective children, they would recreate a warm family cocoon, joined by friends: the best bulwark against overwhelmingly painful memories.

Mira and François got married in 1991.

In the early days, they spoke English together, but Mira soon signed up for French lessons at the Alliance Française. She also hoped to take classes at the Beaux-Arts Academy. For that, she would need to obtain Belgian citizenship. François, who had been scrambling since Mira's arrival to help her with all the paperwork involved in the immigration status, managed through his painter friend Arie Mandelbaum to get a waiver for Mira to take regular classes at the Beaux-Arts Academy in Uccle, where he was director.

Finding a job remained Mira's principle concern, however.

On the kibbutz, she took care of children. What she hoped to do in Belgium was to create an art atelier for children. After a series of classes given to children in the homes of Israeli families who had moved to Brussels, encouraged by their enthusiasm, she brought her dream to fulfillment. Jean-Marie Bertrand found her a locale in Watermael-Boitsfort where she could organize creative workshops for children aged 6 to 10.

Her success was such that she needed to find a larger space, which she did, in the Altitude 100 neighborhood, a wonderful atelier that could hold up to thirty children. "L'atelier de Mira" would operate for eight years to the delight of the children and their parents who were dazzled by the quality of their offspring's art projects.

She needed to work on her own art, however, so in 1999 she signed up for classes at the Saint-Gilles Academy where she practiced drawing, photography, silkscreen printing, lithography, collage, and where she brought together materials into original presentations, sometimes using odd found objects (iconographic documents eaten away by slugs, moldy materials, plant substances, etc.). She graduated in 2009 with a degree and a wealth of solid friendships among both teachers and fellow students.

The walls of the apartment on Rue François Folie are graced with Mira's works, each one both aesthetically pleasing and emotionally charged.

A venture into the unknown, that must be what Mira experienced upon arriving in Belgium. So much to discover and master: a strange little country, a new language, quirky customs and ways of doing things. But perhaps the most exhilarating was the discovery of the world of physics and the community of physicists. She fell under the spell of the warm and welcoming atmosphere that reigned within this community of intelligent, creative, and sharing people. Though physics itself remained beyond her reach, being a part of "François's world" offered her curious mind some choice food for thought. She accompanied François on all his many trips, attended countless scientific gatherings, and shared his enthusiasm. For her, François's friends would soon form a real family that she adopted and who adopt her unreservedly.

Intermezzo: Pleasure, Like a Beacon in the Fog

Once again, François and I were together, seated in the living room. We were no longer in the highly charged apartment where François lived for nearly thirty years, first on his own and then with Mira, but in the apartment where the Englerts lived since 2015 — a spacious, bright apartment surrounded by greenery, with a large terrace that Mira turned into an enchanted garden.

We were trying to do a review of the work we had accomplished thus far.

During our earlier interviews, François was having a lot of trouble speaking about his childhood, recounting his life as a hidden child, those years that his conscious mind seemed to have blocked. But little by little, oftentimes at the price of painful memory work, he had managed to put words to this period of his life. It was this recovered speech that had allowed me to share with readers the emotions of the child he once was.

The lonely child grew into a lonely teenager.

And then, the lonely teenager of the post-war years grew into an adult mindful of others. *Humani nihil a me alienum puto* — Nothing human is alien to me.[8] François's circles of friends included countless artists, artisans, the highly literate, philosophers, book lovers, and lovers of the land, all more or less marginal, for he remained the absolute non-conformist that he had forever been. He loved talking about his friends and delighted in telling innumerable anecdotes, often hysterically funny, that had spiced his life.

Yet, I was still unsatisfied.

For I knew that this man who gave an undeniable impression of fulfillment was still terribly angst-ridden.

—   François, I sometimes get the feeling I'm missing something, that there are memories that have been given the boot.
—   *That's quite possible. You know, back in the day, when I would think about my life, it was as if everything was drowned in a kind of fog. Before the war, during the war, and even afterward. The absence of a past. A lack of reference points. My parents' silence.*
—   You sound like Georges Perec.

And I read him a passage from *Ellis Island* that I noted one day in the margin of the text: "… In some way, I'm estranged from myself. […] I don't speak the language my parents spoke, I share none of the memories they may have had, something that was theirs and made them what they

---

[8] An annotation by Montaigne inspired by a line from the Latin poet Terence.

were — their story, their culture, their hope — was not handed down to me."

— *Yes, that's exactly it. And that plunged me into a kind of double isolation, from my earliest childhood, in both space and time. A hole in my life, a void that I couldn't put words to. As I grew up, it was this void, I believe, that aroused my need for independence, and a certain aggressiveness toward the outside world. I felt an obscure need to exteriorize this aggressiveness, a way to fill the void, and this is most certainly what led to my protestor's turn of mind. By contesting everything, I created excitement, just the outlet I needed. For a long time, I had the impression I was aimlessly wandering, like a boat adrift.*

    But once my research work started to produce increasingly exciting results, I felt a great easing. The boat had finally come into port. I wasn't aware of it back then, but I had finally caught a glimpse of what would from then on be the governing principle of my life. The pleasure of discovery. You talk about it in the text, actually, my discovery of the pleasure of... discovery! The pleasure of seeking and finding, to always push a little further, now that's something I can put words to, that I can share.

— Right, I can see that clearly. But tell me, you've traveled quite a bit, haven't you? Aren't there places that provided you with something? Landscapes, cities, the sea?

— *The Far East, yes, but other than that, not so much.*

— What about painters?

— *Annette Deletaille taught me to appreciate painting. But for all things visual, that's really Mira's world!*

— And music?

— *Ah music, that's something else. At home, Marc and I used to talk about music a lot. I must have been fifteen when Marc had me listen to Bach's violin concerto in E major. I was utterly fascinated. We used to listen avidly to a class in music history given by Robert Wangermée on the radio, the INR. He would play excerpts from major works. I remember the Adagio and Fugue in C minor by Mozart that left a huge impression on me. Chopin was also a revelation. And then Beethoven.*

One day, a former high school friend of mine who had an art shop near Place St.-Jean had me listen to a recording of Dido's lament by Purcell. That was my first contact with classical music for voice.

Years later, Esther and I were digging around the flea market to find that recording. After about forty-five minutes, we found the original 1951 LP with Kirsten Flagstad in the role of Dido. So gorgeous!

I also remember crazy listening sessions with Georges Miedzianagora and Robert De Mayer who had a real ear for music. We were at my place, with a couple bottles of wine and the Bach Mass in B minor. We drank, we listened, and the more we drank, the more beautiful the music sounded to us, such vigor, extraordinarily powerful!

That music holds a special place in François's life, all his friends know this.

It is always a pleasure for me to see him sit down at the piano and play a waltz or a Chopin nocturne. He plays with elegance and intelligence, which is how you have to play this music at once romantic and marvelously structured.

He recently introduced me to a recording of Schubert's Trio in E-flat major for piano and strings with, on violin, Jean Orloff, one of his former students and professor at Clermont-Ferrand, on cello, another friend, and at the piano François Englert!

And one must not forget, poetry, of course.

At the most unexpected moments, François would start reciting or singing, with that lovely deep voice of his, his favorite lines from the poets he loves most. At other moments, it will be a tirade from Moliere.

And many are the streets of Brussels or elsewhere that one night or another have resonated with one of his makeshift recitals.

Episode IX

International Recognition

IX.1. A Cascade of Prizes

In 1997, François Englert, Robert Brout, and Peter Higgs received the **European Physical Society Prize**, awarded in Jerusalem, "for their hypothesis of the existence of a new elementary particle, the boson, responsible for the emergence of non-zero mass particles." Martinus Veltman was the one who submitted Brout and Englert's candidacy, and it was at the KU Leuven that they would be invited to celebrate the event.

A thrill and a surprise for François and Robert, for until then, only Peter Higgs had been consistently cited with regard to the mechanism that all three of the researchers had imagined, and the "Higgs boson" was the term widely in use. Henceforward, the "BEH boson" would be the preferred term.

This first official international recognition of their work heralded a change in the lives of the two researchers, and things would now start to speed up.

Higgs would not accept the award in Jerusalem, for he was boycotting Israel for its policy toward the Palestinians. Robert Brout would make a brief appearance, as his wife was gravely ill and he could not be gone for long.

In fact, François did not meet Higgs in the flesh until 2012, at an unofficial event at CERN. The contact between the more reserved Briton and the "feisty *Bruxellois*" was friendly enough, though somewhat restrained. Brout, on the other hand, had already met with him a few times.

In 2004, Englert, Brout, and Higgs received the **Wolf Prize**, awarded in Israel, the most prestigious prize in physics after the Nobel. At the same award ceremony, the Wolf Prize for the Arts was awarded to the pianist and orchestra conductor Daniel Barenboim, an artist wholeheartedly committed to seeking a peaceful solution to the Israeli–Palestinian conflict.

Despite this concurrence that he might have approved, Peter Higgs once again did not show up. He had written to Robert Brout that he intended to come, but that he no longer could in light of Israel's assassination of Sheikh Yacine! In a chivalrous move, François insisted that he not be eliminated from the award despite his absence, as the rules stipulated.

It was during his participation at a summer institute at the University of Chulalongkorn in Bangkok organized by Anne Taormina that François learnt that the Wolf Prize was to be awarded to none other than Robert Brout, Peter Higgs, and himself. François, who had gone with Mira, Laurent Houart, and Marc Henneaux, met up with Professors Tohru Eguchi, a colleague of Anne's, and Jnanadeva Maharana, whom he knew from CERN. To celebrate the event, the little group decided to go have a big blow-out party on one of those island paradises not far from the city, in a Vietnamese restaurant that also featured Japanese karaoke, at which Mira was a natural. Professor Maharana arrived with a huge bouquet of flowers meant for François, but on the way in, he tripped, and to his great embarrassment, fell flat on his face!

In 2010, François Englert, Robert Brout, and Peter Higgs along with three American physicists, Gerald Guralnik, Carl Hagen, and Thomas Kibble, received the **J.J. Sakurai Prize** awarded by *The American Physical Society* "to honor outstanding achievement in Theoretical Particle Physics." The ceremony took place in Washington DC. One of the traditions pertained to the recipients' acceptance presentation, for which they were allotted a certain amount of time. Higgs was once again absent. Robert attended the event, though already quite sick at that point. His speech had become quite labored and the idea of having to make a little presentation was anxiety-inducing. François asked that he might be granted Robert's time allotment in order to speak on his behalf. The audience would later recall with emotion the warm words that François devoted to his friend, who was visibly overjoyed by what François had to say. In the evening back at the hotel, someone knocked at François's hotel room door. It was Kathy, Robert's second wife, and she was in a panic. Robert was extremely upset, and kept repeating incessantly: 64 and 78, 78, 78! François understood right away: the two dates 1964 and 1978, the article on the BEH mechanism and the one on quantum inflation! Brout greatly valued their scalar boson work, but he felt that their work on Cosmology — their version of Genesis — was their most important contribution to science.

On 4 July 2012, at a public lecture, CERN announced that they had identified, with 99.99997% certainty, a new boson in a mass field of the order of 125 GeV, which manifested instability properties and disintegration features predicted by the BEH mechanism.

People were beginning to say that Belgium might well add a Nobel Prize to its modest collection.

But King Albert II did not need a final confirmation to demonstrate his admiration of our scientist. In October 2012, he invited François Englert to give a talk about his research at the Laeken Palace, as part of a reception where the upper crust of Belgian society would be in attendance. But first, a kind of dress rehearsal was organized, a buffet lunch where François would give his presentation before the scientific advisory board and a certain number of notables and academicians. François did his best to render the complex material accessible to a non-specialist audience who might have trouble visualizing what he was saying. He carefully avoided scrawling equations on the board. But once François had finished, an aide de camp of the King approached: "You know, it might be nice if you could show the King a couple equations. He would like that. Could you show us a couple?"

A few days later came the dinner at the palace, with a photo session and introduction to a whole assembly of learned personalities. There was not a single woman in attendance! Mira was not invited either!

After his lecture, into which François slipped a few equations, Albert II approached François: "So, those equations, they're yours?" "Yes, these equations are mine!" Albert II was beaming. He was friendly, unceremonious, and fun-loving, in stark contrast to the atmosphere of the Belgian royal house reputed for its austerity. He felt a genuine fondness for François, which would translate in July 2013 into his bestowing upon him a grant of hereditary nobility, a kind of knighthood, and the title of baron.

This would prove to be one of his last official acts before his abdication on 21 July 2013.

In May 2013, François Englert, Peter Higgs, and the CERN (in the person of its director Rolf Heuer) received the **Prince of Asturias Award**, the most prestigious Spanish prize, for "the theoretical prediction and experimental detection of the Higgs Boson, a fine example of how Europe has made it possible through a joint effort, to solve one of the most profound mysteries in physics." The award ceremony took place with much pomp and circumstance at the Oviedo Teatro Campoamor and was followed by a grand reception attended by the people of Oviedo. François and Mira were given a chauffeured car that took them to the Guggenheim Museum in Bilbao, opened especially for them.

As for the Nobel ...

Winners of the Nobel Prize are traditionally announced in early October, and this year, it was on 8 October at eleven o'clock in the morning.

It was with a certain feverish nervousness that François, Peter Higgs, and CERN were awaiting the announcement, which is always communicated by phone to each of the winners.

The Swedes are very punctual people, and yet, nothing happened at the fateful hour.

François, Mira, and François's daughters Anne, Sarah, and Hélène had gathered in the Uccle apartment.

There was a brief radio communique, however: the announcement of the Nobel Prize will be an hour late.

And indeed, it was a little past midday when the phone rang, announcing the great news, and minutes later, it was posted to the website nobel-prize.org:

The 2013 Nobel Prize in Physics is awarded to François Englert and Peter Higgs "for the theoretical discovery of a mechanism that contributes to our understanding of the origin of mass of subatomic particles, and which recently was confirmed through the discovery of the predicted fundamental particle, by the ATLAS and CMS experiments at CERN's Large Hadron Collider."

This prize rewarded the theory elaborated by Robert Brout, François Englert, and later Peter Higgs in 1964, which had already garnered a trio of prizes, from the European Physical Society in 1997, the Wolf Prize in 2004, and the Sakurai Prize in 2010, which was experimentally validated in 2012, and later earned François Englert, Peter Higgs, and CERN the Prince of Asturias Award in May 2013.[1]

Why the one-hour delay, François wondered?

Eleventh-hour deliberations, no doubt.

The Physics Prize is awarded by the Royal Swedish Academy of Sciences on a motion by the Nobel Committee which examines the list of nominees. Only those scientific institutions and figures whose status is duly specified are eligible to present candidates: members of the Royal Swedish Academy of Sciences and/or the Nobel Committee, former Nobel Prize winners, and tenured professors at Universities and Institutes of Technology from all over Scandinavia and the Karolinska Institute of Stockholm. Others solicited to nominate candidates are tenured professors from at least six non-Scandinavian universities as well as a number of scientists chosen by the Academy. The Committee's proposal is examined by the Academy's Class for Physics, which can propose modifications. And in the end the Academy decides by majority vote the very day of the announcement.

Since the maximum number for any one prize is three, one might well have expected to see François, Peter Higgs, and CERN. But such will not be the case, only François Englert and Peter Higgs would be the fortunate chosen ones.

The substance of the deliberations is subjected by decree to a fifty-year code of silence.

Will we find out, then, someday far in the future, why the Academy crowned two laureates instead of three? Close scrutiny of the wording, "for the theoretical discovery of a mechanism," might help explain why CERN is missing.

But François likes to think that the absence of a third laureate is perhaps a way of honoring Robert Brout, the co-author, who would surely have been at his side if he had not died in 2011.

---

[1] Robert Brout died in 2011.

François, Nobel Prize winner … First reaction from friends who knew his parents: "Such a pity your father can't be here; he had so much faith in you, he would have been so happy …"

Regrets, yes.

But lots of joy, too. And well-deserved pride as well, for François had never sought to play the influence game among the many important figures who were already on his side, never a hint of lobbying on his own behalf — a practice not entirely unknown in the upper echelons of science.

The official announcement on the radio touched off thunderous applause in the ULB rectory hall on the Solbosch[2] campus where a throng of colleagues had gathered. In Uccle, beneath François's apartment windows, journalists were jostling for a scoop. François did the papal thing, arm raised skyward in an ecumenical salute to the crowd.

- So Professor Englert, are you happy?
- Well, yes! I'm not unhappy, at any rate! It's wonderful to be acknowledged by such a prestigious prize.

Later, in a large hall on the La Plaine campus, the physicist was awaited by administrators, the media, numerous government ministers, and a crowd of researchers, students, and colleagues.

A standing ovation! The hall was all smiles.

François would later aver, "It's the first consequence of this prize, people are so nice to me, I'm a bit surprised. And a little worried, for I fear that could change!"

François first paid tribute to his friend Robert Brout and expressed his immense regret at not being able to share this great moment with him. He next commended the work of Belgian universities and called for unconstrained and adequately subsidized research.

Taking questions from journalists, he responded with humor and good will. No, he could not explain to them the BEH boson in five minutes!

The Prime Minister warmly congratulated François for "this prestigious distinction that crowns one of Belgium's great minds and an exceptional career in the service of science."

By late afternoon, the whole family had gathered at his daughter Sarah's to celebrate the event. Grandson Gilles and granddaughter Olivia,

---

[2]The ULB occupies two campuses, the older one being Solbosch and the newer one La Plaine.

in their great excitement, climbed up on the roof of a car parked in the street and held up a sign in praise of banana toast, because "The Nobel Prize, yes, but also the banana toast prize!"

Later, François would confess to one of the journalists:

"The Nobel was obviously a highlight. In all truth, we had been expecting it. The results obtained at CERN before July 2012 were already strongly hinting at the boson! But I have to admit that the thrill was even greater in 1964 when Robert Brout and I understood that our hypothesis held up under scrutiny. We uncorked some champagne to celebrate that."

IX.2 In the Aftermath of the Nobel Prize

In October 2014, King Philippe implemented François's accession to the title of baron. The new baron is always to choose a motto. His first choice was one of La Rochefoucauld's maxims: "He who lives without folly is not as wise as he thinks." And for his emblem, golden cats! Why cats?

> *Both ardent lovers and austere scholars*
> *Love in their mature years*
> *The strong and gentle cats, pride of the house,*
> *Who like them are sedentary and sensitive to cold.*
>
> *Friends of learning and sensual pleasure,*
> *They seek the silence and the horror of darkness;*
> *Erebus would have used them as his gloomy steeds:*
> *If their pride could let them stoop to bondage.*
>
> *When they dream, they assume the noble attitudes*
> *Of the mighty sphinxes stretched out in solitude,*
> *Who seem to fall into a sleep of endless dreams;*
>
> *Their fertile loins are full of magic sparks,*
> *And particles of gold, like fine grains of sand,*
> *Spangle dimly their mystic eyes.*[3]
>
> Baudelaire, *Les Fleurs du mal*

---

[3] Aggeler, W. trans. "Cats" from *The Flowers of Evil*, Fresno, CA, Academy Library Guild, 1954.

But the rules of heraldry are categorical: the motto must be held by the cat's paws, but the graphic designer pointed out that the La Rochefoucauld maxim is too long. It will therefore read: "No Wisdom without Folly."

One of the metallic balls of Brussel's famous landmark, the Atomium, now bears the name of François Englert!

There is a *Rue François Englert* in Tubize, an *Avenue François Englert*, a *Venelle*[4] *Robert Brout* as well as a *Venelle du Boson* in Uccle,

---

[4] Small alley.

and on the ULB La Plaine campus, a *Chemin François Englert* and an amphitheater named "François Englert and Robert Brout Hall."

A stamp was issued as well as a 5-euro coin, both with the image of the two friends.

In July 2018, François, Peter Higgs, and Fabiola Gianotti, director of CERN, received the **Carla Fendi Prize** in Italy, at Spoleto, in conjunction with the *Festival dei Due Mondi*, or Festival of the Two Worlds. This is a festival held every year since 1958, bringing together a whole range of artists, dancers, actors, filmmakers, puppeteers, writers, painters, and since 1980, scientists as well, for the organizers felt that Science is also an Art. The eclectic François, friend to poets, painters, and musicians, was very touched to receive an arts prize. A short, poignant film showing the important stages of his life commemorated the event.

In 2019, it was finally decided, at the highest echelons, to renovate a beautiful ruin of a chateau in the Tournay-Solvay Park, on the edge of the Soignes Forest, for the purpose of founding, in this place where architecture meets nature, a center of excellence devoted to physics: the BEL Center, for Brout, Englert, and Lemaître (Section VI.2). The plan, sponsored by four universities, ULB–VUB and UCL–KULeuven, was to create a place for sharing, "a Queen Elizabeth Music Chapel for scientists,"[5] in the words of Fadila Laanan, Undersecretary for Scientific Research.

François was invited to visit the premises in the company of architect Francis Metzger and Fadila Laanan. Wearing his red hat purchased in China and his red scarf, he spoke enthusiastically about the future BEL Center. Though he had until then refused nearly all requests to lend his name to projects, he was delighted by this one.

"You'll have to engrave inside the building itself," he says, "an image of the Brout–Englert–Higgs boson, and an image symbolizing Georges Lemaître's major contribution to our understanding of the expanding universe."

The Center would organize seminars in particle physics, to become a permanent international gathering place for students and researchers. It would also offer outreach activities in particle physics and cosmology for a broader public.

---

[5]The Queen Elisabeth Music Chapel is a Belgian academic institution for artistic training of young musicians, which was created by Queen Elisabeth of Belgium in 1939.

François wanted these activities to be extended to other disciplines such as philosophy, for instance. The architect, for his part, spoke of the profound connection between architecture and science. François approved: architecture, science, philosophy, nature, art, all these categories came together in a certain idea of beauty, coherence, and simplicity, but he emphasized how hard it was to make current physics accessible to a broad public. This was why the creation of a center like the BEL took on a particular importance.

On 8 September 2020, the non-profit organization BEL was signed into being by the representatives of the four partner universities, the ULB, the VUB, the UCL, and the KULeuven, in the presence notably of Véronique Halloin, General Secretary of the FNRS, of Barbara Trachte, Undersecretary for the Brussels-Capital Region, and a number of François's colleagues, all duly masked, as pandemic rules required. François ended his brief introduction with these words that rang with truth in the hallowed hall of the old Salle des Marbres, where the wild hopes of the May 68 assemblies still resound:

> "Fundamental research is the quest for a rational understanding of nature. But today, it seems to have lost its powers of enchantment and is even scorned in certain sectors in the West. And yet, this is where it was born, with the revolution of the Renaissance, to become an essential feature of our culture. Its decline could well signal a resurgence of irrational, destructive and dangerous ideologies which, so often in the past, have resulted in barbarity, once again threatening us today with the violence it necessarily entails. This is why it is critical that the BEL Center be not only a success in research, but that it become a vehicle for optimal dissemination of our research findings to all levels of education and learning."

When it comes to the BEH boson itself, the notion most accessible to the public probably remains that highly simplified but meaningful image that François had given of the first seconds of our universe:

> *We can imagine scalar bosons as particles that form a condensate, a sea that envelopes the whole universe. This condensate appeared during a transition phase in the first hundredths of thousandths of a second that*

followed the birth of the universe. Within this sea, certain particles that had no mass acquired mass. This is what allowed the world that we know, and ourselves along with it, to exist.

This image inspired the journalist-physicist Guy Duplat to wax lyrical about what "unifies" arts and sciences, to François's utter delight:

This boson ushers in dizzying new perspectives, on par with the discovery of DNA in biology: the void of our universe is believed in reality to be filled by a huge field, a sea. [...] There are rare moments in the history of mankind: when Bach composed his sonatas, when Rembrandt painted the human mystery. The major scientific discoveries deliver the same emotion. This Nobel Prize arouses a particular joy, that of witnessing how much intelligence and humility before the facts has been committed, how much tenacity, sheer joyous will to pierce the world's secrets and dispel the gloom of ignorance. A choice that is also an encouragement to invest in fundamental research, the kind that, like art, like love, does not demand an immediate yield, for which there is no immediate 'use,' except that it is the very salt and greatness of our life on earth.

[...] François Englert has allowed us to dream! And the Nobel has crowned not only a fundamental discovery that sheds light into the world's darkness, as do all great artists, but this prize also brings to the fore a fiercely free, original and creative man.[6]

IX.3. A Laureate's Journeys

After the awarding of the Nobel Prize, there was of course the first journey, the trip to Stockholm. François's children, and Mira's, their partners and all the grandchildren were along for the occasion.

On 10 December 2013, after the ceremony at the Konserthuset (concert hall), where King Carl XVI Gustaf bestowed upon the laureates their medal and diploma, after the speeches, came the reception and sumptuous banquet at the Stockholm City Hall. The banquet was followed by a ball, but Mira and François were commandeered for a very official photo

---

[6] Baré, Fr. and Duplat, G. *Particules de vie: Conversation avec François Englert*, Brussels, Renaissance du livre, 2014.

session in the presence of the royal family. Next, the younger generation proposed that they join a less ceremonial festivity, a dance party organized by students. Mira and François made a brief appearance, while the younger set danced until dawn.

Exhausted by all these ceremonies, François and Mira yearned for a little getaway to Florida, but an email arrived from Belgium announcing an official reception. To the apology that François sent back, hoping to postpone the event, a return email arrived announcing that a private jet was being sent just to come get them, him and Mira! Goodbye, Florida!

Next, an avalanche of invitations, offering François and Mira a veritable world tour.

The Far East, which François had already had the chance to explore, and which had always held an immense appeal for François: the streets of Bangkok, their vibrant bursts of life, their heady aromas. The people's gaze, intense, curious, amused, never serene, that unabated effervescence, day and night.

China, Japan, Thailand, Singapore, and Vietnam as well.

Not to mention the countless invitations coming from Europe and the Americas, notably from Chile, and of course, from the United States.

One particularly memorable trip was to Rwanda in 2017, on the invitation of Neil Turok, South African cosmologist, founder of AIMS (African Institute for Mathematical Sciences), an excellence network created to enable the best students in Africa to become engines of scientific invention in their countries under the hopeful motto "The next Einstein will be African." On 3 April 2017, President Paul Kagame officially launched in Kigali the "Ecosystem of Knowledge" project whose purpose was to place Rwanda as the African vanguard in scientific research. He then yielded the floor to the guest of honor, François Englert.

François developed the themes closest to his heart.

- *Fundamental research and creativity*: "The impact of fundamental research on society is of course owed to the technical applications derived, directly or indirectly, but is also due to its intrinsic creativity. Without creativity, there is no genuine progress."

- *Fundamental research and the pleasure of knowledge*: "Fundamental research is not only a source of economic development, it fills life with the immense pleasure felt by anyone who acquires knowledge."
- *Rationality and the future of humanity*: "It is crucial to promote fundamental, experimental and theoretical research, and its dissemination at all levels of education. The challenge is to attain a sustainable scientific and technical development and to valorize knowledge and rational thought in order to build a more peaceful, humanist and livable world."

He praised the AIMS network initiative and stressed the importance of coordinating efforts across numerous African countries to foster the enthusiasm that young Africans demonstrated for the sciences.

His speech drew warm applause.

"A moving speech by Professor François Englert, Nobel Prize, on hope and the passion to learn that he has observed in Rwanda and in Africa," one attendee wrote in her blog.

Mira and François next visited the Genocide Memorial, a site that commemorated the 1994 genocide of the Tutsis. The chauffeur made available to them throughout their stay had himself seen his entire family massacred during the weeks of horror. The Memorial, sober and moving, is remarkably designed, and since its inauguration in 2004, thousands of survivors have come to pay respects.

The Memorial also includes a permanent exhibition on the history of genocide in the world, in particular, the Shoah and the Armenian genocide — homage to all the victims. This Memorial is thus a powerful educational tool for generations to come.

Episode X

The Birthday Party

This episode presents a few quick portraits of some of François's friends. Readers unfamiliar with our protagonist's circle might get the impression that this episode does not concern them, but our purpose is twofold: on the one hand, to celebrate this fondness for friendship that so defines François, and on the other hand, to highlight a little-known aspect of the life of physicists for a public that often imagines them as a community of sedentary scientists alone in their labs or offices. Nothing could be further from the truth! They are people who journey, people who share. We invite you here to come along with them, to share for a moment the vibrancy of the life of these wanderers of science.

X.1. The Sixth of November 2012 was François's Eightieth Birthday

The International Solvay Institutes of Brussels[1] marked this occasion, from 5 to 7 November, by organizing a major colloquium entitled: "The Quantum Quest: A Fascinating Journey."

The context surrounding this particular moment was far from trivial: a few months earlier, on 4 July 2012, after a half-century of tumultuous attempts, CERN announced its detection of the BEH boson, the particle predicted by the BEH mechanism, validating this prediction. The CERN experimenters, the distinguished crafters of this discovery, were in attendance at the Colloquium. The idea of a Nobel Prize was in the air, but François was not overly concerned by it, since he had never been one to seek out honors, just as he had nothing to do with organizing the Colloquium.

It was **Laurent Houart's** idea originally, but unfortunately he died during the preparation period. He was one of François's most talented doctoral students, and had become a close friend. A session at the Colloquium was held in tribute to this physicist who passed away far too young.

The title of the Colloquium faithfully reflected Laurent Houart's concern to illustrate François's masterful contribution to the practice of physics: the importance, on the one hand, of the quantum validation of all new theories, and on the other, the close ties between quantum mechanics and cosmology — a fascinating journey indeed between the infinitely small and the infinitely large.

---

[1] It was after the success of the first Solvay Conference in 1911 that the International Solvay Institute for Physics was founded in 1912, and a few years later, the Institute for Chemistry. The mission of these institutes is to promote scientific research to "deepen our knowledge of natural phenomena" by fostering meetings and discussions among researchers.

INTERNATIONAL SOLVAY INSTITUTES
BRUSSELS

THE QUANTUM QUEST: A FASCINATING JOURNEY
TO CELEBRATE FRANCOIS ENGLERT'S 80TH BIRTHDAY

Brussels, 5 - 7 November 2012
Université Libre de Bruxelles - Campus Plaine, Solvay Room

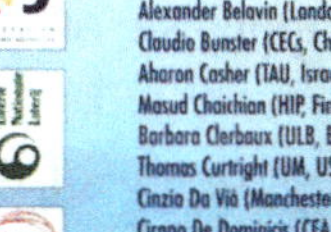

INVITED PARTICIPANTS INCLUDE:

Yakir Aharonov (TAU, Israel)
Luis Alvarez-Gaumé (CERN, Switzerland)
Daniele Amati (ICTP, Italy)
Ignatios Antoniadis (CERN, Switzerland)
Costas Bachas (LPTENS, France)
Máximo Bañados (PUC, Chile)
Alexander Belavin (Landau Institute, Russia)
Claudio Bunster (CECs, Chile)
Aharon Casher (TAU, Israel)
Masud Chaichian (HIP, Finland)
Barbara Clerbaux (ULB, Belgium)
Thomas Curtright (UM, USA)
Cinzia Da Vià (Manchester U, UK)
Cirano De Dominicis (CEA, France)*
Bernard De Wit (Utrecht U, The Netherlands)
Nathalie Deruelle (IHES, France)
David Fairlie (Durham U, UK)
Louis Fayard (Orsay U, France)
Pierre Fayet (LPTENS, France)
Willy Fischler (Texas U, USA)
Peter Freund (Chicago U, USA)
Belen Gavela (UAM, Spain)
Fabiola Gianotti (CERN, Switzerland)
Nigel Glover (Durham U, UK)
Michael Green (Cambridge U, UK)
Gerard 't Hooft (Utrecht U, The Netherlands)
David Horn (TAU, Israel)
Lydia Iconomidou-Fayard (Orsay U, France)
Jean Iliopoulos (LPTENS, France)
Joe Incandela (UCSB, USA)
Joseph Katz (HUJI, Israel)
Costas Kounnas (LPTENS, France)
Maria Krawczyk (Warsaw U, Poland)
Dominique Lambert (FUNDP, Belgium)

Viatcheslav Mukhanov (LMU, Germany)
Shmuel Nussinov (TAU, Israel)
Jean Nuyts (UMons, Belgium)
Jean Orloff (LPC, France)
Renaud Parentani (Orsay U, France)
Giorgio Parisi (Sapienza U, Italy)
Tsvi Piran (HUJI, Israel)
Stefan Pokorski (Warsaw U, Poland)
Sandu Popescu (Bristol U, UK)
Eliezer Rabinovici (HUJI, Israel)
Marianne Rooman (ULB, Belgium)
Augusto Sagnotti (SNS, Italy)
Pierre Sikivie (UF, USA)
Kelly Stelle (Imperial College, UK)
Harry Stern (ULB, Belgium)
Leonard Susskind (Stanford U, USA)
Anne Taormina (Durham U, UK)
Charles Thorn (UF, USA)*
Guido Tonelli (CERN, Switzerland)
Jean Tran Thanh Van (Orsay U, France)
Anca Tureanu (HIP, Finland)
Peter West (King's College, UK)
Paul Windey (LPTHE, France)

ORGANIZING COMMITTEE

Riccardo Argurio (ULB, Belgium)
Jean-Marie Frère (ULB, Belgium)
Marc Henneaux (ULB & Solvay Institutes, Belgium)
Axel Kleinschmidt (Max-Planck-Institut, Germany)
Alexandre Sevrin (VUB & Solvay Institutes, Belgium)
Philippe Spindel (UMons, Belgium)
Michel Tytgat (ULB, Belgium)
Isabelle Van Geet (Solvay Institutes, Belgium)
Antoine Van Proeyen (KUL, Belgium)

SCIENTIFIC COMMITTEE

Riccardo Argurio (ULB, Belgium)
Lars Brink (Chalmers, Sweden)
Jean-Marie Frère (ULB, Belgium)
Marc Henneaux (ULB & Solvay Institutes, Belgium)
Axel Kleinschmidt (Max-Planck-Institut, Germany)
Hermann Nicolai (Max-Planck-Institut, Germany)
Pierre Ramond (UF, USA)
Alexandre Sevrin (VUB & Solvay Institutes, Belgium)
Philippe Spindel (UMons, Belgium)
Michel Tytgat (ULB, Belgium)
Antoine Van Proeyen (KUL, Belgium)

www.solvayinstitutes.be

* To be confirmed

The poster (see opposite) was designed based on a screen print by Mira where two silhouettes, identifiable as François and a friend, are gazing out at a horizon that seems empty, but where there are undoubtedly many hidden secrets not yet revealed by neither the infinitely small nor the infinitely large worlds, and which will be evoked in the Epilogue — something to dream about as you read to the end of this chronicle.

Many of the physicists François had worked with and grown close to were invited, over sixty responded positively to the call, all thrilled to be there, brimming with stories and memories to share. Several guests would address the assembly during the two-day event to talk about François, and especially about physics. The papers presented at the conference, all of the highest quality, reviewed the most cutting-edge research, though always ending with uninhibited birthday wishes, to the wild applause of the assembled participants.

The organizers also took into account the non-scientist friends of the honoree: for them, Mira planned a cultural excursion to Bruges, which would prove a huge success. The visitors were dazzled by the Memling Museum, where the highly skilled Mira initiated them to the wonders of paintings from the Flemish Primitive School.

On the evening of 6 November, a dinner was planned in the grand hall of the Plaza Hotel, for the Colloquium participants, François's family, their children and partners, his brother Marc with his partner Jeanine, and a large number of friends. This was a chance for François to get together with all those dear to him, and for this beautiful art deco hotel, nearly one hundred years old, younger sibling to the Georges V in Paris, a chance to welcome within its venerable walls the cream of modern physics.

The tradition of after-dinner talks would be upheld. **Paul Windey**, a former doctoral student of François's and a close friend, and **Pierre Ramond**, master of ceremonies, took over this light-hearted event — two after-dinner talks full of wit and fun. Pierre Ramond had prepared a little performance: right in the middle of his speech, his cell phone rang, he took the call, his face dropped, and with a crack in his voice, he announced that, alas, a terrible error has just been discovered: it was not the BEH boson that was detected on 4 July! Hushed silence in the hall… an uneasy moment or two, but fairly quickly, the audience got wise and burst out laughing, with applause for the actor.

And since they had all gathered for this memorable birthday dinner, why not seize the occasion and introduce the physicists among the guests?

Their names are all there on the lovely colloquium poster.

We've already made the point that for François, friendship has always been paramount. For him, this eightieth birthday would represent, in a way, the culmination of one of the "fundamental laws" in the Englertian notion of the world: the law of sharing, for the sharing of discoveries, of emotions, of dreams, is for him the sweetest, most uplifting, most fecund thing for scientific work and for life more broadly.

You should know that François is a marvelous storyteller; he excels at evoking moments experienced with his friends, and the countless lively anecdotes he injects into conversation are always a joy to hear. During the hundreds of personal interviews that made the present chronicle possible, the anecdotes poured forth like fine wine. François was especially insistent on recounting the more outlandish scenes. Because with each story, he revives friendships, encounters, and events that have been for him *the salt of the earth and the light of the world* — an array of elements that provided a powerful antidote to the ghosts of his past, and François feels a great sense of gratitude toward all those who shared their trust, their time, their enthusiasm or their craziness. There are so many that it would be wishful thinking to even attempt to cite them all. But all were important and hold a special place in François's memory. You, the readers, have already met a few of them in the interstitial sections entitled "L'école buissonnière," and several others throughout Episode VII.

But now that we have addressed François's "law of sharing," the locations of such sharing take on critical importance. There are those inevitable venues such as colloquia, conferences, and summer institutes that occupy a researcher's agenda, where ideas face off, where affinities are discovered, and where colleagues' work-filled days usually end with convivial meals and get-togethers. And then, in the four corners of the globe, there are those university departments where François has always been welcomed with open arms. All these places form a grand constellation where researchers converge, in a cycle of meeting and dispersing according to a ramified timeline. We propose that readers accompany a few of them, set out on what may prove a perilous enterprise as the threads of their journeys get tangled. To avoid getting lost in this maze, we will organize our list according to locations, following them from place to place … and an index of names is available at the back, for the readers' convenience.

Among the friends who took part in the 80th birthday colloquium that day, before taking their seats around the grand round tables at the Plaza, there was one in particular who played a major role in organizing meetings among physicists:

Jean Tran Thanh Van, a sweetheart of a man, professor at the Université d'Orsay. This portrait can be found on the blog of the alumni of Lycée Jean-Jacques Rousseau in Saigon: "If physics is his life's work, international exchanges constitute his realm, and his field of research is quite simply extraordinary." And from the 4 August 2000 edition of *FermiNews*:

> "Some people really know how to throw a conference. The acknowledged maestro, the dean, the Pearl Mesta of physics conferences is a diminutive Vietnamese theorist from the University of Paris at Orsay. He is the founder and organizer of three of the most respected scientific get-togethers going, the Rencontres de Moriond, the Rencontres de Blois, and the Rencontres du Vietnam in Hanoï. Beginning with Moriond in 1966, these annual events have inspired and excited generations of physicists, experimentalists and theorists alike, […] rendering a creative and sustained service to particle physics, international understanding, and the humane aspects of science."

In 2013, this indefatigable near-octogenarian created in Quy Nhon, along with his biologist spouse Kim, an International Conference Center, "to make Vietnam a rallying point for sciences […] because the emerging countries need fundamental science. Without basic research, the world cannot advance. All the revolutions we are experiencing spring from the quest for knowledge without prior assumptions, research that does not obtain immediate answers and does not seek instant profits."[2] Both strive

---

[2] Interview by Dominique Leglu in *Sciences et Avenir*, 28 June 2016. https://www.sciencesetavenir.fr/decouvrir/agenda/jean-tran-thanh-van-sans-science-fondamentale-le-monde-ne-peut-pas-avancer_23494.

to make the teaching of science more accessible to Vietnamese children and teens, in collaboration with **Jean-Marie Frère**, current director of Theoretical Physics at the ULB, in charge of the theory program at the Rencontres de Moriond.

François participated several times at the Rencontres de Moriond where CERN regularly announced its latest discoveries, as well as at the Rencontres de Blois. Where the Blois gathering was considered more prestigious, taking place in the sumptuous setting of the Château de Blois, the atmosphere at Moriond always felt more like family, and the meetings among experimenters and theoreticians, young researchers and their elders, have always been a delight for François.

For our little world tour of places where science and so many other things have been shared, what more natural place to begin than CERN, with its experimenters and theoreticians, many of whom would become François's friends:

Fabiola Gianotti, spokesperson in 2012 for ATLAS, the giant detector at CERN in Switzerland, a radiant, warm, and enthusiastic woman. In 2016, she would become the first woman director of CERN. In 2018, she would receive the Fendi Prize along with Peter Higgs and François Englert.

Joe Incandela, spokesperson in 2012 of the CMS,[3] the other large detector located in France, remarkable for the aesthetic concern that seemed to have governed its construction, and whose colored spires of its immense electromagnets call to mind the petals of amazing flowers. It was Joe Incandela who would announce with Fabiola Gianotti, on 4 July 2012, the discovery of the BEH boson, an item that made front-page news the world over.

Guido Tonelli, also employed at CERN, and another key figure behind the detection of the famous boson. Spokesperson of CMS in 2011, he announced the first promising collisions and kept François abreast of events by telephone. He predicted that the existence of the BEH boson would be confirmed (or not) by the end of 2012, perhaps earlier, for "the CMS teams have obtained an unbelievably rich yield of new physical

---

[3] ATLAS and CMS are two of the four collision points of the Large Hadron Collider (LHC), CERN's gigantic particle accelerator/collider.

measurements with extraordinary results beyond our most optimistic expectations."

The Spaniard **Luis Alvarez Gaumé**, physicist and excellent pianist, and his Italian wife **Cinzia da Via**, both active at CERN for many years. This was where François met Luis and Cinzia, and again in 2006 at a colloquium organized in Friesland to mark the sixtieth birthday of **Gerard 't Hooft** (Section V.4). On the way back, in Amsterdam, François, Mira, Luis, and Cinzia dined at a Japanese restaurant where Cinzia, a glass of sake in hand, launched into a laudatory speech about François. The four friends discovered many penchants in common. François has an enchanted memory of the musical *Mary Poppins* which he saw in Manchester with Cinzia. Luis and Cinzia were present the following year when François received the Asturias Prize and during the sumptuous party that followed at the Oviedo Teatro Campoamor. They both work presently at Stony Brook University in New York.

Maria Krawczyk, of the University of Warsaw, who was particularly interested in the BEH mechanism. During her stays at CERN, she organized workshops on the topic. Once the LHC was in place, she would focus on the physics of this particle accelerator. In 2004, she sat on the programming committee for the Rencontres de Moriond. She invited François and Mira to Poland and provided François with an office at the university. It was there that François met **Stefan Pokorski**, director of the Department of Theoretical Physics in Warsaw, who had spent long residencies at CERN that, in his words, had a decisive impact on his research career. One day, Stefan Pokorski knocked on François's office door in Warsaw. He started telling him bits and pieces of his life. As it happened, Stefan was also a hidden child. It was in the woods where his mother had taken refuge that she gave birth to him. The mother and child remained hidden in the woods for a long time, surviving miraculously thanks to assistance by the father who was not Jewish. With that, he invited François to stop by to sample some pierogis. And as was often the case on their trips to Poland and Russia, Mira set out in search of vestiges of her family, but in Warsaw, nothing remained; the family home located in the ghetto had disappeared.

Let us leave Switzerland for Italy, home to meeting centers that are quite popular among researchers, notably the International Center for Theoretical Physics in Trieste, referred to in Episode VII, as well as the Ettore Majorana Foundation and Center for Scientific Culture[4] in Erice (Sicily)

The Erice Center, located in a set of former monastery buildings, had been organizing every year since 1963 a series of workshops and seminars in various scientific disciplines. It had become a renowned meeting place for scientists the world over, welcoming such luminaries as Paul Dirac, Murray Gell-Mann, Yoichiro Nambu, Abdus Salam, Sheldon Glashow, Peter Higgs, and of course François Englert and Robert Brout. The Center was headed by Professor Antonio Zichichi, a great friend of the Pope, assisted since 1998 at the Subnuclear Physics Section by **Gerard 't Hooft**, who had become the co-director. In that capacity, he had the great privilege of lodging in what was called the "papal" chamber, which he showed to François, who recalls the sumptuous bed whose size was anything but monastic!

Antonio Zichichi was directing the summer schools with a firm hand. Oversight of the students was strict and the system for obtaining scholarships very competitive. Zichichi personally kept watch over the students' every move to determine which of them was actively participating and which of them was not (the so-called "ghosts"). One readily imagines that however much François admired the high standard of the classes, he was less appreciative of the atmosphere. In 1999, François and Robert Brout were included in the program, on the theme of "Basics and Highlights in Fundamental Physics," and on other occasions, François met up with **Hermann Nicolai** and **Bernard de Wit.**

Let us cross the Atlantic now for the "winter schools" at the University of Florida (VII) where François would have the great pleasure of getting together with **Thomas Curtright** and JoAnn, **Pierre Ramond**, and **Kelly Stelle** from Imperial College, the Swede **Larks Brink** from Göteborg, a pioneer in Superstring Theory, and his wife Åsa; all became

---

[4] Named after the physicist born in 1906, known for his work in particle physics, and who disappeared under mysterious circumstances in 1938.

inseparable during these winter sessions that were more festive than studious. Lars Brink and Pierre Ramond were both part of the Colloquium Board.

Let us not forget the warm welcome in South Carolina where the University of South Carolina at Columbia, a public university, has proven particularly generous toward physicists. Every year, it offered the group from Tel Aviv residencies with all the facilities and space necessary for their work. The three musketeers, **Yakir Aharonov**, associate professor at Columbia since 1973, **Roni Casher,** and **Shmuel Nussinov** didn't have to be asked twice, nor did François, the fourth musketeer.

Heading to Asia now, to discover the summer school at the University of Chulalongkorn in Bangkok, organized in 2004 by **Anne Taormina** and her former doctoral student **Auttakit Chattaraputi.**

The students were a diverse group: Europeans, Thais, some of whom were Anne's students at Durham, Japanese, and Vietnamese, the latter standing out from the other more carefree participants by their serious and rather stuffy demeanor, a little "Germanic" in François's words. François, Mira, **Laurent Houart,** and **Marc Henneaux** got together with many friends there, with whom they would roam through Bangkok, a city François was already a bit familiar with. He was once again fascinated by the hustle and bustle in the streets, the colors, the aromas, the people. All were surprised at the impressive number of opticians, dentists, tailors, and of course massage parlors. François and Mira left later with Laurent Houart to visit the Khmer temples of Phimai, a three-day trip organized by a local agency: a lot of sightseeing and gastronomical adventures.

Anne Taormina is a longtime friend of François's. In February 1984, François attended her dissertation defense at the University of Mons, under the direction of **Jean Nuyts**, also present at the Colloquium. It was during her doctoral work that she got increasingly interested in elementary particle physics. François and Anne met again when she started working for CERN, and later did a postdoc there in 1987. They would remain in regular contact, based as much on their common interests in physics as on their great mutual trust. It was at CERN that Anne would meet her future husband, the physicist **Nigel Glover**, professor at the University of Durham, a historic town in northwest England where the university was housed in an impressive 11th century castle. Anne was quick to land a teaching position there. Next, she had a visiting professorship at the University of Chicago before returning to Durham and becoming head of the Department of Mathematics where, in addition to working, like François, "in pursuit of the unification of the four fundamental forces of nature," she endeavored to promote the participation of women in university research. Every time she went back home, she would stop and visit François. In England, they traveled together through the county of Durham, still a somewhat wild area that shares the same natural beauty as neighboring Scotland, and visited a number of medieval castles, all haunted, of course. Anne told François her story, happy for such a sympathetic ear. Her life was far from simple: a high-powered career, a husband, two children, passions, nostalgias. One day, on one of their outings, the two friends bought sandwiches and decided to eat them in the car, a very British sort of station wagon. They settled into the back seats, not thinking about the child-resistant door locks. They couldn't get out! The front seats were unreachable, the windows didn't open! Not a soul in sight. Time went on, and what they thought was a funny misadventure started feeling less so. The two friends were finally rescued by a handful of passersby on their way to the pub, who had a good laugh as they took pity on these *bloody foreigners*.

In 2010, Nigel Glover took the initiative to organize an international meeting in Durham on the BEH mechanism. François, Robert Brout, and Peter Higgs were among the first to be invited. But fate had other ideas: a few weeks before the scheduled date, the Icelandic volcano Eyjafjöll threw a fit. The eruption sent huge clouds of smoke and ash into the European sky, making air travel impossible, and the meeting had to be cancelled. One of the rare occasions for François, Robert, and Peter to "talk shop" thus fell through. To mark the event, Anne had commissioned a multicolored plate of Venetian glass representing the famous "Mexican sombrero," a three-dimensional image describing the BEH boson. Today, the plate hangs on a wall of François and Mira's apartment.

And now, on the road to the Middle East and of course the University of Tel Aviv and its musketeers, **Aharon Casher** and **Shmuel Nussinov** (the third, **Yakir Aharonov,** was unfortunately absent) where François always felt welcome.

Hebrew University of Jerusalem was the workplace of his old friend **Joseph Katz** who had come to celebrate François with his wife Ruthi, and his brilliant doctoral student **Tsvi Piran**.[5] It was also the home university of **Eliezer Rabinovici**, director of the Israeli Institute of Advanced Studies, who had also come to Brussels for the birthday party. This renowned physicist had worked not only for the development of Israeli science but also for better understanding between Israelis and Palestinians. He had a dream: to reconcile the peoples of Israel and Palestine through scientific collaboration, to make science a bridge between the two peoples. In 1999, thanks to the solidarity of men and women of good will and under the aegis of UNESCO, Arab and Israeli associations, the Palestinian Authority, Jordan, Turkey, CERN, the European Union, and even Iran and Pakistan, the idea was born to provide the Middle East with an international laboratory on the CERN model. It was called SESAME,[6] located in Jordan, a great step toward realizing his dream. SESAME, CERN's younger brother, is equipped with a particle accelerator. Scientists from the world over work there, but many are from the Middle East. As Rabinovici put it, *They are scientists who have broken the deadlock; [...] science is a universal language that evolves on the margins of politics. When you enter here, you leave the conflicts behind.* And the dream continues. *We have begun with science, and we shall build other bridges.*[7] The Center was officially inaugurated in 2017, and in September 2021, Eliezer Rabinovici was elected President of the CERN Advisory Board.

It was at the University of Jerusalem that François met up with the Russian cosmologist **Viatcheslav Mukhanov**, with whom Eliezer Rabinovici had written some articles. A graduate of the Moscow Institute of Physics and Technology, he was posted at the Ludwig Maximilian University of Munich, the alma mater of Werner Heisenberg and Max

---

[5] His dissertation, *Astrophysical Processes near Black Holes*, 1976, was directed by Joseph Katz and Jacob Shaham, the father of violinist Gil Shaham.

[6] Synchrotron light for Experimental Science and Applications in the Middle East.

[7] *Québec Sciences*, 31 March 2014.

Planck. An exuberant fellow, lover of good wines, he had generated important advances in the study of the origin of the galaxies. In 2015, **Tohru Eguchi** of the Universities of Tokyo and Chicago would organize in Sendai, Japan, a session of the Gravity and Cosmology Workshop. He would invite François who, in turn, invited **Mukhanov** and **Philippe Spindel**. The following year, at a conference in Copenhagen, it was Mukhanov who would invite François. They would get together often, notably in Israel and Brussels.

Back in Old Europe, the École Normale Supérieure of Paris is another of the world's preeminent gathering places, where François met up with **Jean Iliopoulos**, graduate of the University of Athens, one of the founders of the Theoretical Physics Laboratory at the École, author of several books on elementary particles, one of which was prefaced by his friend François with whom he had an especially close friendship and who came to appreciate his talents as chef and wine connoisseur.

Among the guests, on that 6 November 2012 day of celebration, some were less connected than others to this or that gathering place, but rather were friend who had grown out of casual encounters:

Nathalie Deruelle, director of research at the CNRS was a great friend of Joseph Katz, Tsvi Piran, and Philippe Spindel. An intrepid traveler, she collaborated notably with the Yukawa Institute of Theoretical Physics in Kyoto. During one of his work residencies in Paris, François stayed at Nathalie's, at Rue Gît-le-Coeur, one of the oldest streets in Paris, famous in the 1950s among the Beat Generation, since authors Allen Ginsberg and William Burroughs lived there for a time. François was charmed!

Jean Orloff did his doctorate with Robert Brout and François, and *remembers fondly discussions with François and Robert that allowed him to appreciate the extraordinary complementarity of their approaches to physics.*[8] Today, professor at Blaise Pascal University in Clermont-Ferrand, an excellent violinist in his spare time, he was the one who initiated François to chamber music.

---

[8] Interview by Laurent Sacco in *Futura-Sciences*, 18 October 2013. https://www.futura-sciences.com/sciences/actualites/physique-boson-brout-englert-higgs-cosmologie-49545/.

Masud Chaichian, a Judeo-Iranian physicist, came in from far-off Finland, which he had fallen in love with, accompanied by **Anca Tureanu**, his charming Romanian wife, also a physicist. Masud was the first "non-native" professor appointed at the University of Helsinki, the culminating point of a career that had taken him from Moscow to Rehovot in Israel, from England to Germany, then to CERN and finally to Finland. François was always welcome at Masud's Finnish home, provided he removed his shoes upon entering, a hard-and-fast rule in all of Scandinavia, which struck François as odd! Masud even offered him a pair of slippers to coax him into abiding by the rule.

Several guests have come especially from the USA, Germany, Chile, and England:

Peter Freund from the University of Chicago, one of the pioneers of String Theory. A Jew born in Romania, he barely escaped the Shoah and came close to being shot after an anti-Soviet demonstration in 1956. He managed to leave Romania in 1959. Also known as a novelist, he had written novels and short stories published online in the *Exquisite Corpse* series. He saw numerous parallels between literature and science: *most papers in physics are short stories, in which concepts, rather than human characters, undergo adventures.*[9]

In 2004 at a conference, **Axel Kleinschmidt**, of the University of Cambridge, met Marc Henneaux, Laurent Houart, and François. This marked the beginning of a collaboration with the Theoretical Physics Department at the ULB, where he would eventually be hired in 2007, and also the start of a great friendship with François. Axel Kleinschmidt then became a professor at the Max Planck Institute. François often visited him at his home in Golm, a neighborhood of Potsdam where the Max Planck Institute for Gravitational Physics was located. Like Hermann Nicolai, he was a member of the scientific board for this present Colloquium.

A few of the friends Laurent Houart made at the University of Santiago in Chile, where he was a regular visiting researcher, attended this Colloquium, notably the great Chilean physicist **Claudio Bunster**, director of the Santiago Center for Scientific Studies.

---

[9] https://news.uchicago.edu/story/peter-freund-particle-physicist-and-fiction-writer-1936-2018.

While no one could have predicted that he would one day become a physicist, Bunster became a pioneer in high-flying theoretical physics in Chile. It was during his high school days that he was allured by the "magical" sound of the word *fisica*, and decided to learn physics all by himself. After graduating with honors from the University of Chile and Princeton, he embarked upon a career marked by his non-conformist streak. Professor at Princeton and University of Texas, he returned to Chile at a time when most intellectuals were looking for ways to get out of the country, then under the boot of the bloody dictator Augusto Pinochet. Banned from universities by the military government, he managed against all odds to found an autonomous research institute, one that went against the flow of mentalities and subsidies. *I felt that at this precise moment, I could be more useful, and that the best way to do so was to set up an independent location where colleagues from all over the world could come bring some light into a gloomy situation.* During the presidency of Eduardo Frey, he headed the Scientific Advisory Board and played a capital role in reconciling civil society and the army. He aspired to consolidate democracy through science by involving the army in scientific projects. These improbable collaborations would result in several extremely productive expeditions to the Antarctic.

In 1978 while he was a visiting fellow at Princeton, **Marc Henneaux**,[10] a young researcher at the ULB, first met **Claudio Bunster**. This marked the start of a long collaboration, notably on the issue of black holes. Henneaux considered this physicist, eight years his senior, his spiritual father and a sustained collaboration between Claudio and the ULB took shape.

Claudio Bunster was a multifaceted character. He loved to have fun, throw parties, and he enjoyed shocking people for its own sake.

Such a free-wheeling personality would necessarily win over someone like François. They would often get together, whether in Chile or Belgium. They would frequently be in league when one or the other played some practical joke, something they both adored. One day, during

---

[10] Professor at the ULB and the College de France. When they were in Belgium, he and his wife Eli gladly respected the tradition of going to eat mussels at Kriekske, in Lembeek, in the company of François and Mira.

a celebration in honor of Marc Henneaux on one of the countless islands off the Chilean coast, Claudio suggested to François, who was to be introduced to some officials, to turn toward him once the introductions were over and shout a resounding "Fuck you!" François accepted on the condition that Claudio responded with an equally resounding "Fuck you, too!" And they did it! After a stunned silence, those in attendance burst out laughing as fireworks lit up the Chilean sky.

And finally, across the Channel, a salute to **Peter West** of Kings College London. **Abdus Salam** was his dissertation director. Peter West created a research group on supersymmetry and string theory at Kings College. He held concurrent positions as visiting professor at CERN, at Stony Brook, at the California Institute of Technology, and other prestigious research centers. A kind, warm-hearted man, an expert in M-Theory.[11] He showed up one day at François's in Brussels to propose that François, Laurent Houart, and Anne Taormina work together on the symmetries of M-theories: *You are the perfect team for the job*! After a few months, they decided to write an article summarizing their findings. Peter West wrote a first draft in which he analyzed the whole history of the M-theory. The others thought it was too long, and discussions got heated over the phone. The article would finally get published in a version slightly abridged by Anne Taormina, who would be sharply rebuked for this sacrilegious act.

And then, there were all the local friends, and among them, those who labored to organize this Colloquium sponsored by the Solvay Institutes. They came from our universities: ULB, VUB, FUNDP, UMons, and KULeuven. Their names were also on the poster.

Among them were **Harry Stern**, collaborator and longtime friend of François and Robert, **Riccardo Argurio** and **Marianne Rooman**, both former doctoral students of François. In the early 1980s, François had proposed that Marianne devote her doctoral dissertation to elaborating an approach to the 7-sphere, drawing upon octonion algebra,[12] a research field off the beaten path that Marianne had never dealt with. Starting

---

[11] Theory popularized by Edward Witten, which supposedly unifies the superstring and supergravity theories.

[12] Octonions are a generalization of the notion of number. (Episode VII)

completely from scratch, Marianne was able to make a masterful contribution that would be published in *Nuclear Physics*.[13] Despite what looked like a promising career in particle physics, Marianne later chose to move into the study of various fundamental problems in biology. Today, she is a research director at the FNRS and a world-renowned specialist in Bioinformatics.

We should also cite **Antoine van Proeyen**, member of the General Assembly of the BEL project (Section IX.2). He had known François since he participated in 1986 at the NATO Advanced Research Workshop on Super Field Theories in Vancouver, where **Lars Brink** also took part.

Let's end with special appreciation for **Isabelle Van Geet** and **Dominique Bogaerts**, who handled the administrative business of the Solvay Institutes with great efficiency. Whenever François had occasion to visit the Institute, he never failed to stop by to say hello to these charming ladies and share coffee and cakes.

But scientists, whether accompanied or not by their "better half," weren't the only ones seated around the Plaza's large round tables. Among these guests were François's family, his brother Marc and his partner Jeanine, the family of Laurent Houart, whose memory was honored during the Colloquium, and many friends, including Kathy, Robert Brout's second wife, and the charming Marina Solvay, the great-great-granddaughter of the creator of the Solvay Conference (Section III.2), a great friend of Ben Tursch, the accomplice in the Sputnik episode. François fondly recalls the time spent with Marina, at her home or at his.

I was also there, and I will never forget the warm, elated atmosphere that prevailed that evening, from start to finish. There were songs, emotional speeches, funny stories, hugs and laughter, and a few private conversations that snuck in a quantum tirade or two, somewhat dented by the wine, the fine food, and the joy of the gathering.

---

[13] "11-dimensional Supergravity and the Octonionic 7-Sphere," Doctoral dissertation, ULB 1984. Rooman, M. "11-dimensional Supergravity and Octonions," in *Nuclear Physics*, vol. B236, 1984, pp. 501–521.

Epilogue: Glimpses of the Future

There are more things in heaven and earth, Horatio,
Than are dreamt of in your philosophy.

William Shakespeare, Hamlet

When, on 8 October 2013, the Nobel Foundation announced that "Englert and Higgs" were the happy winners of the Nobel Prize in Physics, François received numerous congratulations via email. Among them was a hearty message from Gerard 't Hooft, whose wishes are worth any solemn speech:

Great! My daughter said on her Facebook page: Now the lives of Mr. Englert and Mr. Higgs (both of whom she met several times) will never be the same as before! She knows this is so.

And so it was, just as she said.

There was notably that guest world tour which we summarized in Section IX. 3.

François had always been attracted by the peoples and countries of the Far East, which he was able to experience firsthand when he was invited there to lecture or attend conferences. After receiving the Nobel, his invitations were now in a different class altogether, much to his liking, since he had grown more appreciative of comfort as he moved into his octogenarian years. In Thailand, Vietnam, China, Singapore, and Japan (perhaps more

239

reservedly), he had discovered to his surprise a population that he felt was extraordinarily close to his way of being and thinking, contrary to what many people had reported. He realized how much the wonder he experienced at science's contribution to understanding things could be shared.

This is why he suggested that we end our chronicle with a look at the scientific intelligibility of the world today, with a discussion of the problems it raises that necessarily refer to the philosophical questions humans have grappled with since the dawn of time.

Starting with the Galilean revolution, it took but five centuries for the most astonishing intellectual pursuit in the history of humanity — the search for a rational explanation for all phenomena, whatever they may be — to demonstrate that most phenomena likely to affect us are manifestations of general laws. These laws, integral parts of the "standard model" (Section V.4), govern the basic constituents of all objects, be they inanimate or living, earthly or within our galaxy and the known universe. These basic building blocks, the elementary particles, not unlike Lego bricks, have constructed everything the known universe contains by interacting with each other according to the quantum laws of the standard model. Moreover, space and time, which form the substratum of the universe, developed according to the laws of Einstein's theory of gravitation, i.e., classical relativity, beginning with an initial state born about 13.8 billion years ago.

And yet, as we have mentioned several times throughout this chronicle, we have here two major actors in the construction of the universe that sometimes have trouble agreeing: on the one hand, classical physics that describes excellently the macroscopic world, and on the other hand, quantum physics that describes just as satisfactorily the subatomic world,[1] but where classical physics no longer applies, even though the macroscopic world and the subatomic world are of course "cut from the same cloth!" The classical laws have to be "quantized," i.e., modified to become compatible with quantum physics, but here is where an apparently insurmountable problem arises: to this day, it appears impossible to reformulate the gravitation theory, as classically described by general relativity, in

---

[1]Including certain macroscopic phenomena that only quantum physics can explain, such as superconductivity and superfluidity.

quantum physics, i.e., at the subatomic level, as is done with electromagnetism, where Maxwell's theory has been reformulated as quantum electrodynamics. We can, however, reasonably estimate the scale below which this classical theory of gravitation becomes invalid: it is the Planck scale, i.e., a distance much smaller than the distances encountered in the subatomic world (Section VI.1).

This impossibility plays a decisive role in the fact that the birth of the universe is only partially understood at the current time, and this limitation to our rational understanding of phenomena implies a more general limitation to their intelligibility.

Let's lend an ear to François's thoughts on this ultimately troubling question, during one of our recent dialogues (the italics denote François's part):

I much approve Einstein's vision: "The most incomprehensible thing about the Universe is that it is comprehensible"[2]. *Still, so many things continue to escape our grasp and the "incomprehensible comprehension" of the universe is indeed a particular example of the current limitations of physics.*

Scientists sometimes state the idea that the understanding of "Everything" is already near at hand and that a few minor breakthroughs in our knowledge would suffice so that physics would soon need only to analyze new or already collected findings. I think this is an illusion, one that reemerges periodically in the history of science.

Of course, there are certainly questions that can be answered within the scope of current knowledge or in a (relatively) simple extension of it. For example, it is conceivable that the standard model, which has been validated within the range of currently accessible energies, could be generalized when we explore substantially higher energies. Such generalizations have already been posited, but to this day remain difficult to verify. For instance, it could be that current theories get inserted into a larger Yang-Mills theory, the "great unification," which would unify at very high energy via a BEH-like mechanism the weak and electromagnetic interactions with strong interactions to result in a more unified vision of elementary particles.

---

[2] "Das Unverständlichste am Universum ist im Grunde, dass wir es verstehen können."

On the other hand, this mysterious thing called dark matter,[3] detectable through its influence on the evolution of celestial bodies, is perhaps the manifestation of the existence of new, unidentified particles, stable, neutral, and coupled almost exclusively to gravitational forces, and which would complement the set of known particles.

As for the so-called dark energy,[4] it produces an exponential expansion of the universe. This accelerated expansion is thought to be analogous to the primordial inflation introduced in "Genesis according to Brout–Englert–Gunzig–Spindel," but at a very different scale: this phenomenon currently has weak though appreciable cosmological effects, but it could become dominant in some distant future, if it turned out to be the manifestation of a parameter of general relativity already envisioned by Einstein and labeled "cosmological constant," which describes a repulsive (!) component of gravitation.

A repulsive gravitation? Geez …

Yes, but let's acknowledge that this exponential expansion is not yet well understood. The solution to certain of these problems could call upon new concepts as compared to those that underpin current physics, and this is certainly the case for problems linked to the first instants of the universe.

And what about the Big Bang?

That's just it. The initial singularity of the Big Bang Theory is by definition unfathomable, because a singularity eludes any analysis in physics. The possibility of replacing this singularity with a quantum fluctuation of the void at the Planck scale, which we posited, necessarily involves quantum effects of gravitation that we are yet to know. Even though we can reasonably discount them at larger scales where the standard model plays a decisive

---

[3] It was discovered around 1970 that ordinary matter made up of particles of the standard model represents only 20% of the matter in the universe, the remainder being what is called "dark matter," about which we know next to nothing, but whose existence was suspected as early as 1930 via observation of how galaxies behave.

[4] Dark energy, which represents about 70% of known energy in the universe, is what drives the acceleration of the expanding of the universe observed in 1998 thanks to a new generation of satellites and telescopes.

role, the fact remains that we do not know these effects, and this reduces all current theories about the origins of the universe to the rank of hypothesis.

We are thus faced with a fundamental dilemma: quantum physics applied to matter, on the one hand, and classical general relativity on the other are both experimentally verified with astounding precision. But at least one of these two theories is necessarily incomplete or incorrect.

A lot is being said about "string theory".

Yes, many physicists think that general relativity should be supplanted by the so-called string theory which can be formulated directly into quantum physics. This theory attempts to describe elementary particles, including the hypothetical graviton of general relativity, as excitation of strings. I don't intend to discuss this theory here. All you need to know is that, in principle, it allows for a quantization of gravitation and of the matter coupled with it, but there is no current conclusive experimental verification. That being said, the possibility of constructing a theory of gravitation compatible with quantum physics should not be ruled out.

Then, should quantum physics perhaps be reimagined?

So far, quantum physics has suffered no failures in its experimental verifications. But we do have to own up to the fact that it raises a major conceptual problem.

Its "indeterminate" side?

That's right. As we have seen, quantum physics introduces a probabilistic element into the way we should comprehend phenomena. For centuries, physics (of the macroscopic world) was dominated by the conviction that nature obeyed deterministic laws.

In everyday life, our resorting to probabilities results from the practical impossibility of predicting the result of an experiment because of how many and how complex the relevant causes are. In general, this way of going about things has proven legitimate.

For instance, a six-sided dice rolled "at random" will generally land, on average after a large number of rolls, one time out of six on a given side, e.g., the three. We therefore attribute to each possible result a one out of six probability, which translates the average probable occurrence of this result after a large number of repeated rolls.

More generally, in statistical physics, they analyze phenomena that involve a large number of very small objects. They are interested, for instance, in the macroscopic properties of a gas such as its pressure and temperature. It is of course impossible to know accurately the position and speed of each atom or molecule making up the gas, but it will suffice to be able to calculate a distribution of the probabilities of different variables, i.e., their statistical distribution.[5]

Conversely, quantum physics introduces for the result of certain measurements an a priori probability. Quantum indeterminacy appears irreducible since it stems from probabilities that don't rely on a set of causes.

But is that really the case? What do you yourself think?

There's a lot at stake here! In the application of quantum laws, the existence of an irreducible indeterminacy opens the door to lots of speculation! Einstein was nevertheless very reluctant when it came to this possibility; as a proponent of Spinoza's determinism, he thought that quantum physics was incomplete,[6] and that to complete it, there had to exist a deterministic underworld that would cure physics of this indeterminacy where "God would roll dice," which offended him greatly.[7]

---

[5] Statistics is the field of mathematics that collects, processes, and interprets a dataset in terms of probabilities.

[6] In 1926, he wrote to his friend Max Born, "Dear Born, quantum mechanics is absolutely worthy of respect. But a voice in my head tells me that it is not yet the rare find. This theory contributes a great deal, but it fails to bring us any closer to the secrets of the Old Guy. At any rate, I'm convinced that 'He' doesn't roll dice."

[7] As suggested in the famous quip that he made to Niels Bohr during one of their many discussions during the 1927 Solvay Conference. The incident was reported by Bohr: "On his side, Einstein mockingly asked us whether we could really believe that the providential authorities took recourse to dice-playing ("… ob der liebe Gott würfelt"), to which I replied by pointing at the great caution, already called for by ancient thinkers, in ascribing attributes to Providence in every-day language." Schlipp P.A., *Albert Einstein, Philosopher-Scientist*, Evanston, Library of Living Philosophers, 1949.

For Bohr, however, indeterminacy was intrinsic to quantum physics and he saw there the basis of his vision of a somewhat mysterious "complementarity" concept that I will not develop here.

Right, since the famous Solvay Conference of 1927, physicists are at odds on the subject. Just like the philosophers who have been fighting it out for millennia. For as soon as you say "determinism," you think inevitably of its rival, "free will," necessarily conjuring Democritus, Aristotle, Lucretius, Saint Augustine, Maimonides, and then Erasmus, Descartes, Spinoza, Kant …

We'll come back to that. But I would rather talk first about how I am inspired by the work of physicists who have attempted to objectivize to the greatest extent the problem of quantum indeterminacy: in mathematical terms, by resorting to thought experiments, by elaborating theories, and more recently through extremely ingenious lab experiments. Among them, Gerard 't Hooft, Louis Vervoort, Carl Brans notably, have done nothing short of exploring the viability of a deterministic quantum physics.[8] Gerard 't Hooft in particular suggested totally original paths of research.

All these works feature an enormous subtlety, and I'm going to try here to make some of their results comprehensible, though that will necessarily involve some over-simplification.

In the sixties, the Irish physicist John Steward Bell was studying the possibility of defining an underworld in quantum physics consisting of parameters called hidden variables, governed by deterministic laws, and which would ascribe to probabilities in quantum physics the status of statistical mechanics of a deterministic underworld, thereby eliminating the irreducible indeterminacy of quantum physics.[9]

---

[8] Brans, C. "Bell's Theorem does not Eliminate Fully Causal Hidden Variables," in *International Journal of Theoretical Physics*, vol. 27, no. 2, 1988, pp. 219–226; Vervoort, L. "Bell's Theorem: Two Neglected Solutions," in *Foundations of Physics*, vol. 43, 2012, pp. 769–791; 'T Hooft, G. "The Cellular Automaton Interpretation of Quantum Mechanics," in *Fundamental Theories of Physics,* Heidelberg, Springer-Verlag, 2016 (Section 3.6); 'T Hooft G. "Ontology in Quantum Mechanics," in *arXiv. quant-ph 2107.1491v1* (29 July 2021) (see also references cited in the article https://arxiv.org/pdf/2107.14191.pdf).

[9] Bell, J.S. "On the Einstein Podolsky Rosen Paradox," in *Physics*, vol. 1, no. 3, 1964, pp. 195 290.

So how would that work?

We're going to test out this idea of hidden deterministic variables in a case that will seem to you a little remote from the classical vision. And we shall see how this brings us to call into question some fundamentals of physics, and philosophy as well.

Are you ready?

Ready and willing!

Great. Let's consider the experiment where an atom disintegrates by emitting two photons that move away from their source (the residual atom) along a same rectilinear line but in opposite directions.[10] In this scenario, the quantum state of the two photons is called quantum entanglement or EPR.[11] The photons will both have the same polarization direction,[12] but in an EPR state, it is impossible to specify this polarization direction.

The photons end up at polarizers A and B, which have been prepared, let's say, by Alice and Bob, to present respectively the passage directions **a** *and* **b***. Even though the polarization direction has not been specified, it is still possible in quantum physics to calculate the four conjoined quantum probabilities of passage (or of reflection) in the polarizers A and B for each of our two photons.[13]*

If the experiment diagrammed above is repeated a large number of times, we observe that the values of these conjoined quantum probabilities, identified as mean experimental values, are verified.

At this point, we have to generate, according to Einstein's idea, a deterministic underworld that would "legitimate" the probabilistic character of the results obtained.

Precisely. But before I go any further on this venture, let's recall here a fundamental principle of physics, emerging from Einsteinian relativity, the

---

[10] Alain Aspect, "Présentation 'naïve' des inégalités de Bell." Institut d'Optique Théorique et Appliquée–Centre universitaire d'Orsay–avril 2004 arXiv : quant-ph/0402001.

[11] For Einstein, Podolski, and Rosen.

[12] For notions of polarization, see Section III.3, where they are defined.

[13] Passage-passage, passage-reflection, reflection-passage, reflection-reflection.

principle of locality, whereby a signal cannot propagate at a speed beyond the speed of light. In the kind of experiment we're dealing with here, where two photons move away from their source in opposite directions, this means that each photon goes toward the proximate polarizer, without any information issuing from the distant polarizer being able to reach it.

Are you still with me?

Got it, keep going.

Let's go back now to our search for a deterministic underworld. To do that, let's introduce into our entangled system, via a thought experiment, some deterministic hidden variables, whose conjoined statistical probabilities we hope will be equal to the conjoined quantum probabilities we calculated earlier.

We can, for example, take as deterministic hidden variables some well-defined polarizations for the pair of photons and then repeat the experiment a great number of times. In his famous theorem issued in 1964, John Stuart Bell demonstrates a particular property of the classical probabilities then obtained (the so-called Bell's Inequalities), on the assumption that there exists no dependency between the orientations of the polarizers and the values of the hidden variables; in other words, on the assumption that there is independence between the measurement device and the variables measured, what is generally called "independence of measurements."

Sounds perfectly natural to me, as a hypothesis.

Yes, but wait; life is full of surprises.

At any rate, for the moment, in the experimental scheme under consideration, the locality condition is assumed to be satisfied, as well as the independence of measurements, since the orientations of the polarizers are freely regulated by Alice and Bob. But here's the thing: by the end of his demonstration, Bell finds that the quantum probabilities, as opposed to the statistical probabilities, do not respect the "Bell's inequalities" announced above! Impossible to obtain equality between the statistical probabilities of a local hidden variables theory and the quantum probabilities. Impossible to envision local hidden variables that would make it possible to interpret quantum probabilities as statistical probabilities of a deterministic underworld.

And impossible to cure quantum physics of its "indeterminacy."

Maybe not. Because there is a way to render inoperative the Bell theorem by invalidating one of its hypotheses, the one that you found so natural. We can, while maintaining the locality, violate the independence of measurements, and in that case, the Bell theorem would no longer oppose a deterministic quantum physics.

Yes, but the independence of measurements, here, springs from the free choice to regulate the polarizers' orientations, like a kind of free will with Alice and Bob, right? How can that kind of thing be violated?

You have to realize that the notion of free will is so rooted in our societies that, for years, the Bell theorem was almost never questioned. But things changed by and by. In 1988, notably, Carl Brans, of Loyola University in New Orleans, proposed that we back up a little — literally. I'll try to summarize what he did.[14] *It's clear that in a deterministic world, each event is located along a causal chain, and if we keep going back in time, we will quickly arrive at an event that will contain within its causal future the whole Bell experiment. In other words, the deterministic hidden variables, which "saw" the photons emitted by the disintegrating atom, can now "see" the entire preparation of the experiment, including that of the polarizers by Alice and Bob; hence, it is not unreasonable to think that the polarizers are regulated, not freely, but by a phenomenon analyzable in terms of physics. And we might include — why not, after all — their orientations **a** and **b** among the deterministic hidden variables, which until now had defined only the polarizations of photons. Analysis of the experiment with the help of these hidden variables shows that there will necessarily be a dependency between the measurements and the polarizations — the locality being conserved, nonetheless. Independence of measurements therefore does not exist, and even though the locality is conserved, the Bell theorem can no longer be applied. When the experiment is repeated a great*

---

[14] The initial state of the system analyzed by Brans is different from the state of the two entangled photons under consideration in this epilogue. Applying his reasoning to the initial state of our two photons changes nothing of the essential results he obtained, but allows the reader to understand the pertinence of his analysis without having to assimilate new concepts.

number of times, the results show, for the polarizations, a statistic distribution for each a and b couple, and Brans argued that the classical statistic probabilities would be identical to the conjoined quantum probabilities.

We could have called this Epilogue "Escapes into the Past" ...
I'll be needing a little time to digest all this, you know ...

As I suspected, but what matters is that this result suggests that we can envision a deterministic quantum physics capable of concurring with the fundamental laws of classical physics which are, let's remember, deterministic. In other words, we can conceive of a universe governed by determinism. Which implies that quantum physics, in its current state, cannot claim the status of fundamental science, but would rather be, like so many other theories, a phenomenological theory.[15] "Real" deterministic quantum physics will undoubtedly be difficult to construct, but it seems to me that there should no longer be any theoretical objection to its existence ... so long as this search for a deterministic underworld keeps its promises. Because, however much the hypothesis of hidden variables opens the way to interesting investigation, it does not necessarily define the shape that deterministic quantum physics will take ... if indeed it really exists, which remains to be discovered.

But what about Alice and Bob's free will, in the end?

We were talking about free choice and free will, but is choice ever truly free, aren't we really talking about an illusion due to the immense complexity of the causes that affect an act considered as voluntary?

At this point of the analysis, it might be useful to recall the origins of the scientific interpretation of the world, going back to the Renaissance. When Galileo established the principle of inertia, which provides the basis of how we formulate today's physics, he clearly evidenced that this principle applied to any object, animate or inanimate, living or otherwise. And it is natural to extend this field of application to all of classical deterministic physics.

---

[15] A theory that, without explaining the totality of phenomena under consideration, provides accurate predictions for a subset of these phenomena.

The deterministic laws of physics of course apply to how a body functions, and in particular to the functioning of our some 100 billion neurons and to their vast network of connections. This is the seat of thought and consciousness that necessarily acts in accordance with these laws. It should also be noted that this conception of mine goes beyond traditional materialism to the extent that it rejects matter–thought dualism. Just as we don't believe that there is a color–electromagnetic wave dualism because we know that a (pure) color is a monochromatic electromagnetic wave frequency, and is nothing more, I believe that thought and consciousness are material configurations, and nothing more.

Let's now suppose that the experimental arrangement were effectively determined both by causes exterior to Alice and Bob, and by psychological and physiological mechanisms governing their decision to choose this or that polarization direction. That would eliminate free will!

But you, a free man, can you accept that free will does not exist? *Que le libre-arbitre, ça n'existe pas?*

"ça n'existe pas" …

> *Une fourmi de dix-huit mètres*
> *Avec un chapeau sur la tête*
> *Ça n'existe pas, ça n'existe pas*
> *Une fourmi traînant un char*
> *Plein de pingouins et de canards*
> *Ça n'existe pas, ça n'existe pas*
> *Une fourmi parlant français*
> *Parlant latin et javanais*
> *Ça n'existe pas, ça n'existe pas*
> *Eh ! Pourquoi pas?*[16]

Do you know this little Desnos rhyme?
Oh yes, I love Desnos.

---

[16] An eight-meter ant/With a hat on its head/It can't exist, it can't/An ant pulling a truck/Full of penguins and a duck/It can't exist, it can't/A French-speaking ant/Quoting Voltaire and Kant/It can't exist, it can't/Well, why not? Desnos, R. *Chantefables et Chantefleurs,* Paris, Éd. Gründ, 1952.

That said, there is no free will if we allow for a totally deterministic nature. It is the complexity and the sheer number of causes, and the practical impossibility of knowing all of them that explains the illusion of free choice.

Alright, but will we ever arrive at decrypting the causes, someday? Given their complexity, whether or not they are subject to strict determinism, they will never be absolutely knowable to us. Isn't that a kind of pledge right there of free will?

No, I think that we have reason to hope that this grand question will find precise answers as scientific investigation moves forward, particularly in neuroscience.

And how would you describe this deterministic universe where there is no place for free will?

What I am going to posit now is very speculative and based moreover on currently unverifiable hypotheses.

In a universe governed by deterministic laws, the causes that affect any event are all located in its causal past, which is to say, in the region of space and time from which they can reach the event via signals propagating at a speed that does not surpass the speed of light.

Taking into account that each cause is the effect of at least one previous cause, we can retrace the chain of causes until the origin of the universe, often associated with the Big Bang singularity, if in fact it ever existed, which I doubt ... It seems more reasonable to me to imagine that the universe emerged from the bosom of a quantum "mother" still cloaked in mystery, where space and time were devoid of any operational meaning, but which, at the Planck scale, gave birth to a primordial inflation by means of the BEGS mechanism. This primordial inflation ended at a very high temperature to join the commonly accepted expansion of the universe. Thus, the universe likely emerged with the notions of its time and its space and the notion of the determinism that would govern its deployment.

I won't get into the issue of multiverses here, about which there is no credible hypothesis with regard to their eventual interaction with ours.

As for our supposedly deterministic universe, it is thus from its inception that all the causal chains emerge, giving rise to each event. Such a universe should be understood as a world where all events are connected to each other and are determined by its inception. In this context, the analysis of the experiment linking polarizations and polarizers that we have been discussing, where the role of determinism in the disappearance of free will is manifest, finds its proper place. Of course, to elaborate the theory of entanglement of causes appears insurmountable at first glance, but the problem posed is nevertheless extraordinarily simple metaphysically if indeed there exists a universal determinism that lends itself to scientific analysis. By all accounts, the possibility of developing such a vision requires first and foremost the existence of a deterministic quantum physics which, as I pointed out earlier, has yet to be proven, by a long shot. Our current non-deterministic quantum physics is used successfully, let's be reminded, by all physicists, and is considered as accurate by most of them. Even so, the way it is formulated leaves room for doubt.[17]

I think that a deterministic quantum physics would open the way to a new, coherent vision of the universe, and that it is important to develop this field of research.

You see, ever since Galileo, we have been making impressive progress in knowledge. But the journey is far from over. Among the problems stemming directly from the current state of our understanding of the world, the most important are the elaboration of a quantized general relativity and a deterministic quantum physics. A better understanding of the origin of the universe could be decisive to attaining this end.

So, lots of good work in store.

Yes, and furthermore, we have seen that these problems raise questions that have for a long time remained the purview of philosophy: determinism and free will, thought and consciousness, the human condition, in the end. Okay, how about getting up and stretching our legs a bit?

The building where François and Mira live is surrounded by a large park full of some outstanding trees, already leafing out after the too mild winter of 2021. Many were already flowering. In the little ornamental

---

[17] Bell, J.S. "Against Measurements," in A.I. Miller (ed.), *Sixty-two Years of Uncertainty. Historical, Philosophical, and Physical Inquiries into the Foundations of Quantum Mechanics*, Heidelberg, Springer, 1990, pp. 17–31.

pond, the ducks greeted us; they seemed well acquainted with François who stops by regularly to catch up on their latest.

Our long dialogue continued to resonate in a corner of my head.

These steps in search of determinism proposed by François open up exciting perspectives, but they are ultimately quite natural. Determinism is an age-old concept. Determinism, and its rival, free will, have forever been at the center of countless debates among philosophers, scientists, and clergy of all persuasions.

In the fifth century BCE, Democritus already affirmed that
"Everything that exists in the universe is the fruit of Necessity."
Spinoza, some 2000 years later,[18] said,
"Nothing in the universe is contingent, but all things are conditioned to exist and operate in a particular manner by the necessity of the divine nature." (Prop. 29)
"Men believe themselves to be free, simply because they are conscious of their actions, and unconscious of the causes whereby those actions are determined." (Appendix)
And Einstein, in 1923, speaking of Lucretius, another brother in determinism, said,
"The firm confidence that Lucretius, as a faithful disciple of Democritus and Epicurus, places in the intelligibility, in other words, in the causal connectedness of everything that happens in the world, must make a profound impression."

This round of quotations could go on forever, in one direction or the other, determinism or free will, chance or necessity, these are issues that will remain contentious. For these differing conceptions of the world exert their influence in so many areas, affecting the behavior of physical bodies as much as they do the behavior of living things, and inevitably lead to the moral issue of free will.

How is one to live as a responsible being in a totally deterministic world?

How do we ward off the temptation of fatalism?

---

[18] Spinoza, B. *Ethics*. trans., R.H.M. Elwes. Project Gutenberg, The Project Gutenberg E-text of The Ethics, by Benedict de Spinoza.

Spinoza devoted one of his fundamental works to this question, *Ethics*, published soon after his death in 1677 and banned the following year.[19] It is not an easy read, but what an irresistible encouragement to thinking, so many discoveries, and surprises too, not the least of which are, coming from this austere man, the celebration of the happiness that knowledge brings and the affirmation of a tremendous trust in human intelligence.

I leave readers to meditate upon all this. With that, we can only urge them to follow the counsel of another great François, François Rabelais, an anticlerical cleric, a free-thinking Christian, a stoic, and a bon vivant:

"Therefore is it that you must open the book, and seriously consider the matter treated in it. [...] you must, by a sedulous lecture, and frequent meditation, break the bone, and suck out the marrow."[20]

François Rabelais, Five Books of The Lives, Heroic Deeds and Sayings Of Gargantua And His Son Pantagruel, 1534. (Translated between 1653 to 1694)

Our Chronicle Comes to a Close ...

You, our readers, have journeyed a while in the life of François Englert, physicist, humanist, lover of poetry and song, a man of sharing and friendship.

If this text has had the good fortune of pleasing you and also of arousing your curiosity for the world of physics and its explorers, for its challenges to our thinking and the visionary product of those challenges, we will have accomplished our mission.

François Englert continues to work, to read, to interrogate the nature of things. With the absolute simplicity of a free and curious mind, still weaving it all into the fine work of a lifetime, whose weft is friendship and wonderment inspired by great minds, past and present, be they physicists, poets or philosophers, artisans or musicians.

---

[19] Spinoza was frequently and violently attacked as an atheist during his lifetime. In 1674, on a motion by the provincial synods, the Court of Holland banned the publication of his Treatise on Theology and Politics which he had attempted to publish anonymously in the hope of eluding censorship. Again in 1675, Spinoza attempted to get his *Ethics* published, but it was just after his death that his friends managed to do so. The work would be banned the following year by the secular authorities.

[20] *Gargantua*, from the Prologue. Trans., Thomas Urquhart and Peter Antony Motteux.

Original Texts of the Citations

Episode I, Intermezzo 1

Albert Einstein

Raffiniert ist Herr Gott, aber boshaft ist er nicht.[1]»

Das Wort Gott ist für mich nichts als Ausdruck und Produkt menschlicher Schwächen, die Bibel eine Sammlung ehrwürdiger aber doch reichlich primitiver Legenden. Keine noch so feinsinnige Auslegung kann (für mich) etwas daran ändern.[2]

Epilogue

Einstein's letter to his friend Max Born

Lieber Born, Die Quantenmechanik ist sehr achtung-gebietend. Aber eine innere Stimme sagt mir, daß das doch nicht der wahre Jakob ist. Die Theorie liefert viel, aber dem Geheimnis des Alten bringt sie uns kaum näher. Jedenfalls bin ich überzeugt, daß "der" nicht würfelt.[3]*

Einstein about Lucretius

Einen tiefen Eindruck muß das feste Vertrauen erwecken, das Lukrez als treuer Schüler Demokrits und Epikurs in die Verständlichkeit, bezw. den kausalen Zusammenhang alles Weltgeschehens setzt.[4]

---

[1] [https://einsteinpapers.press.princeton.edu/vol12-doc/52].

[2] [https://de.richarddawkins.net/articles/der-einstein-gutkind-brief-mit-transkript-und-englischer-ubersetzung].

[3] [https://oe1.orf.at/artikel/211752/Ein-Brieffreund-namens-Einstein].

[4] Préface à l'édition bilingue latin-allemand, traduction de H. Diels, 1923.

François Rabelais

« C'est pourquoi il fault ouvrir le livre et soigneusement peser ce que y est déduict. [...] Puis, par curieuse leçon et méditation fréquente, rompre l'os, et sugcer la substantificque mœlle, [...]. »

François Rabelais, La vie très horrifique du grand Gargantua, père de Pantagruel, *1534.*

Bibliography

Baré, Françoise and Duplat, Guy, *Particules de vie: Conversation avec François Englert,* Bruxelles, Renaissance du livre, 2014.

Cohen-Tannoudji, Gilles and Spiro, Michel, *Le boson et le chapeau mexicain*. Un nouveau grand récit de l'univers, Paris, Gallimard (Coll. "folio-essais"), 2013.

Deruelle, Nathalie, *De Pythagore à Einstein, tout est nombre. La relativité générale. 25 siècles d'histoire,* Paris, Belin (Coll. "Bibliothèque scientifique"), 2015.

Iliopoulos, Jean, *Aux origines de la masse: Particules élémentaires et symétries fondamentales,* Les Ulis, EDP Sciences (Coll. "Une introduction à"), 2015.

Luminet, Jean-Pierre, *L'Écume de l'espace-temps,* Paris, Odile Jacob, 2020.

Pais, Abraham, Subtle is the Lord ... *The Science and the Life of Albert Einstein,* Oxford, Oxford University Press, 1982.

Solvay, Marina and d'Oultremont, Catherine, *Fantaisies quantiques. Dans les coulisses des grandes découvertes du XXe siècle,* Paris, Éd. Saint-Simon, 2020.

Sonnert, Gerhard and Holton Gerald, *What Happened to the Children Who Fled Nazi Persecution,* New York, MacMillan, 2006.

Made in the USA
Monee, IL
07 July 2026

56553671R00164